SOLVING THE INTERNET JURISDICTION PUZZLE

Solving the Internet Jurisdiction Puzzle

DAN JERKER B. SVANTESSON

OXFORD
UNIVERSITY PRESS

OXFORD
UNIVERSITY PRESS

Great Clarendon Street, Oxford, OX2 6DP,
United Kingdom

Oxford University Press is a department of the University of Oxford.
It furthers the University's objective of excellence in research, scholarship,
and education by publishing worldwide. Oxford is a registered trade mark of
Oxford University Press in the UK and in certain other countries

First Edition published in 2017

Published in the United States of America by Oxford University Press
198 Madison Avenue, New York, NY 10016, United States of America

British Library Cataloguing in Publication Data
Data available

Library of Congress Control Number: 2017945336

ISBN 978–0–19–879567–4

To Felix, Freja, and Bianca

Foreword

As I write this foreword, a powerful giant of the online world is smarting from the sting of the Canadian judiciary. In a 9-2 majority judgment handed down just over a month ago, the Supreme Court of Canada rejected an appeal by Google Inc. against a lower court's decision to issue an interlocutory injunction enjoining the search-engine operator from displaying, in *any* of its search results, links to a particular enterprise's websites that are being used to breach intellectual property rights.[1] In other words, Google is being required by the judiciary in one country to de-index certain websites from its search-engine platform across *all* countries. Unsurprisingly, Google has just made a counter-move by filing suit in the US District Court for Northern California requesting non-enforcement of the injunction because it allegedly breaches US law and public policy. The procedural acrobatics are reminiscent of last decade's famous transatlantic dispute arising from a French court ordering Yahoo! to prevent Nazi items from being sold on its online auction site.

While intriguing to follow, such acrobatics are expensive and their results frequently messy. At the same time, they highlight some of the contours of what Professor Svantesson politely sums up as 'the internet jurisdiction puzzle'. I write 'politely' because internet jurisdiction could equally, if not better, be described as a quagmire. It has been a quagmire ever since the end of the 1990s, not long after the internet became a ubiquitous backbone for electronic transactions. But the deleterious effects of the quagmire are increasingly felt as globalization, marching hand-in-hand with digitalization, continues to compress the world.

Extending the memorable phrasing by Michael Geist and Judge Nancy Gertner,[2] the nub of the problem is that where there is no obvious 'there there', laws easily get spread everywhere. And although the push to apply national or regional laws across the globe is often paved with good intentions, it increases the potential for cross-jurisdictional clashes, regulatory overreaching, and disruption to communication at the same time as aggravating risk and uncertainty for businesses, consumers, and governments. Disagreement and befuddlement over how to resolve inter-legal conflict exacerbate these problems.

[1] *Google Inc.* v. *Equustek Solutions Inc.* (2017) SCC 34.

[2] Michael Geist, 'Is There a There There? Towards Greater Certainty for Internet Jurisdiction' (2001) 16 Berkeley Tech L J 1345, 1346.

The regulatory quandaries of internet jurisdiction are accordingly in dire need of level-headed, fresh, yet informed thinking. This book provides a solid dose of such thought. It overhauls old dogmas that have hamstrung the sensible resolution of internet jurisdiction issues and it proposes a multi-faceted framework for resolving these issues in a way that is more in tune with the realities of today. Territoriality-based norms are duly singled out for special critique: a central remit of the book is to stamp out the vestiges of a mindset that privileges territoriality as the primary criterion for determining jurisdiction.

As Professor Svantesson implicitly recognizes, the conundrum at hand resembles neither Rubik's Cube nor a traditional jigsaw puzzle. Creating an optimal regulatory framework for internet jurisdiction does not rest on the application of preset algorithms governed by fixed, bright-line criteria; rather, it must rest on a relatively free-ranging, reflexive, and open-ended consideration of multiple factors that is highly contextual. Furthermore, the result will probably not be free of friction: in contrast to a traditional jigsaw, the various elements making up the best solution to a particular dispute over internet jurisdiction are unlikely to fit together in seamless harmony; some of them will likely grate with each other (thus resembling the jigsaw solution engineered by Professor Svantesson's daughter, Freja—more on that in the Preface!), although they will grate less than other solutions. Hence, the optimal resolution of the internet jurisdiction puzzle promises neither a high degree of legal certainty nor a complete absence of tension. But this is hardly unique for internet jurisdiction—most sensibly functioning regulatory frameworks suffer from similar 'imperfections'. In the latter respect, we must also remember the age-old aphorism about the perfect being the enemy of the good.

This book builds on and consolidates an extensive body of scholarship that Professor Svantesson has generated over the past fifteen years or so. I have followed the development of his scholarship closely, right from the time he commenced his doctoral studies, and I am deeply impressed. As a perusal of the bibliography to this book attests, he is extremely productive. Moreover, he is exceptionally prescient, as shown by his seminal work well over ten years ago on the legal implications of geo-location technologies. His breadth of knowledge and interests is huge, spanning far more than internet jurisdiction matters *per se*. This too is evidenced in the following pages, where Professor Svantesson deftly touches upon an array of fundamental jurisprudential issues concerning, *inter alia*, the basic nature of law. More importantly, he not only has a honed ability to parse and criticize existing regulatory frameworks, but to suggest templates for their improvement. And in doing so, his scholarship is imbued with a dogged commitment to principle-based templates tempered by a healthy concern for what is practical. Again, this welcome synthesis of principles and pragmatism manifests itself in the following pages.

All up, this book firmly cements Professor Svantesson's status as one of the most exciting scholars working on regulatory issues pertaining to the online world.

Lee A. Bygrave
Oslo, August 2017

Preface

Ernest Hemingway is said to have made the point that a book should be a new beginning through which the author tries again for something that is beyond attainment; that the author should always try either (a) for something that has never been done or (b) for something that others have tried and failed. Doing so, Hemingway suggests, a lucky author may sometimes succeed.[1]

In this book, I am trying both for something that has never been done, and for something that others have tried and failed at. In fact, I also pursue something that must be seen as a third alternative complementing the two advanced by Hemingway—I here try for something that *I* have elsewhere tried and failed at.

Anyhow, it is probably not a good idea to devote the opening lines of a preface to disclosing things about the book that are bound to disappoint the reader. However, to avoid accusations of misleading and deceptive conduct, I want—in addition to the above—to stress here that this is a book *about* solving the Internet jurisdiction puzzle. And as a book about gardening does not do the gardening for you, and as a book about the German baking tradition does not put *Brötchen* on your table, this book does not contain any all-encompassing ready-made pre-cooked 'solution' as such. In fact, I would argue that it would be premature to try to write a book providing *the* solution to solving the Internet jurisdiction puzzle at this stage. Instead, the more modest aim of this book is to provide the recipe and (some of) the ingredients we will need to be in the position one day to solve the Internet jurisdiction puzzle.

So what is it then, more specifically, that this book is about? And why refer to the topic as a 'puzzle'? Readers of this book will no doubt have a reasonably clear understanding of the term 'jurisdiction'; and what we mean by the 'Internet' requires no discussion.[2] But when the two words are combined, it cannot be assumed that we all speak of the same thing. Several questions

[1] <https://www.brainyquote.com/quotes/authors/e/ernest_hemingway.html>, accessed 8 April 2017.

[2] For a fascinating account of the history of the Internet, see Jon Bing, 'Building Cyberspace: A Brief History of Internet', in Lee A Bygrave and Jon Bing (eds), *Internet Governance: Infrastructure and Institutions* (OUP 2009) 8–47.

may arise, such as when is a jurisdictional claim Internet-related, and when is it not?

For the purpose of this book, I take a broad view of the meaning of 'Internet jurisdiction'. Importantly, as discussed in more detail throughout this book, both public and private international aspects of Internet jurisdiction fall within the scope. As far as private international law is concerned, I take such a broad view that I include not only the question of jurisdiction, as such, but also the related matters of applicable law, the possibilities of declining jurisdiction, and the matters of recognition and enforcement. In my view, these matters form a system in relation to which it is not fruitful to focus on one and ignore the others. And in public international terms, I include all traditional aspects of jurisdiction (prescriptive jurisdiction, adjudicative jurisdiction, and enforcement jurisdiction) as well as what we may call 'investigative jurisdiction'.

As to the use of the word 'puzzle' to describe the challenge to be solved, I have chosen to use it as I think it very well illustrates the method we need to apply. We have before us several pieces that can be used to address the challenges we face in the context of Internet jurisdiction: pieces such as 'comity', 'reciprocity', 'minimum contact tests', 'targeting tests', and 'geo-location technologies'. These pieces need to be combined in an appropriate manner—like the pieces of a puzzle—in order to form the picture we envisage as the future landscape of Internet jurisdiction.

The puzzle metaphor is also useful to illustrate another aspect of the tasks that lie ahead. When speaking on Internet jurisdiction issues at various occasions around the world, I have often concluded by showing a PowerPoint slide with a picture of a few pieces of a puzzle that my daughter Freja had put together at the time she was about 2 years old.

While the pieces clearly were not intended to be put together in the manner my daughter did, they do make an interesting and quite beautiful picture. Speaking to the PowerPoint displaying this 'incorrect' picture, I make the point that we—in the field of Internet law, and perhaps in law more broadly—often have too many, or at least too strong, preconceived notions of what our solutions ought to look like; based on what we have done in the past, we think we know what our future solutions should look like.

I then contrast this to how Freja ignored the picture on the box to the puzzle and how she thus had no preconceived notions of how the puzzle should be put together. And with determination (and perhaps a degree of violence) she then managed to assemble the puzzle's pieces so as to create a much more interesting picture than the makers of the puzzle had intended. I normally then conclude the discussion by making the point that we can learn from this (apart from the violence bit, I suppose) in our pursuit of solving legal puzzles;

after all, as useful as it may be to draw upon previous experiences, we often seem to limit our creativity to choose only amongst earlier solutions instead of engaging in more of a free search for the 'best' solution. So, in writing this book, I have not limited myself to preconceived notions of what the solution may look like; I have sought to re-examine the existing pieces of the Internet jurisdiction puzzle and, perhaps more importantly, to introduce a few new pieces that may help us one day to successfully complete it. But doing so gives rise to an important question; that is, are we really 'allowed' to approach the problem in the manner I have done? This brings us to a central matter for any law reform, but a matter that is particularly potent when it comes to Internet jurisdiction.

Very often, when discussions turn to the topic of Internet jurisdiction, the reaction—almost a knee-jerk response—is for people to exclaim that Internet jurisdiction is a 'difficult' (or, indeed, 'too difficult') issue for there to be any prospect of solutions. The obvious, but seemingly always ignored, question is of course what we mean when we assert that Internet jurisdiction is a 'difficult' issue? If we refer to the fact that people have so strongly held differing opinions of how a solution should look that agreement is unlikely, then there is hope. And surely what is needed then is debate and exactly the type of creativity I advocated earlier. However, if the reason we think that Internet jurisdiction is a 'difficult' issue is found in a perception that the current law—the law we are seeking to reform—is so entrenched that we cannot in fact hope to change it, then perhaps we may as well raise the white flag, surrender, and move on to something productive.

In the text to come I seek to overcome both these difficulties. As to the former—the differing opinions—I present a framework for jurisdiction that builds on well-established principles, and I am (perhaps naively) hopeful that this framework can be a rallying point that is comfortable for a diverse range of interests. As to the latter—the perception that the law on jurisdiction is too entrenched for us to change it—I go to great lengths to show that our current paradigm of jurisdictional thinking does not in fact rest on any particularly solid legal ground and that we actually are free to undertake the much-needed reform.

Acknowledgements

Drawing upon the Australian custom of opening all major events by acknowledging the traditional owners of the land on which the event takes place, I have adopted the habit of using the Acknowledgements of books I write to recognise the authors of relevant works that precede the work at hand. After all, advances in knowledge are rarely (if ever) sparked in isolation, rather they are the result of the author's interaction with existing thoughts, theories, and knowledge, and are unlikely to be achieved in the absence of such external influences.

This book was written during an intensive period from the end of 2016 to mid-April 2017, and it is written with the desire to be current as of 21 April 2017. However, many of the ideas put forward here have been in various stages of fermentation during the four-year research project I have been able to run under the Australian Research Council's Future Fellowship arrangement. In fact, some of the proposals date back to my very first publications on Internet jurisdiction in 2001.

The fact that the ideas I discuss here have developed over a long period of time means that I am indebted to a very large number of people who, in one way or another, put me in the position to write this book. For a variety of reasons, I am not able to include the names of all those to whom I owe gratitude. Some names I have forgotten, others—such as those who have anonymously reviewed articles I have written and the proposal for this book—I just do not know the names of. At any rate, I take the opportunity to thank the following people for their kind assistance, support, co-operation, feedback, and/or comradery: Agata Jaroszek, Alana Maurushat, Alex Mills, Alexander Peukert, Alexander Seger, Anabela Susana Sousa Gonçalves, Andreas Wiebe, Angela Daly, Anna Zeiter, Anna-Maria Osula, Axel Metzger, Benedetta Ubertazzi, Benedicte Ghanassia, Bertrand de La Chapelle, Björn Johansson Heigis, Bob Fox, Brendan van Alsenoy, Carrie Fox, Cathrin Bauer-Bulst, Cecilia Magnusson Sjöberg, Cedric Ryngaert, Chris Kuner, Chris Reed, Christiaan Baardman, Christiane Wendehorst, Christina Schulman, Christophe Bernasconi, Christopher Millard, Christopher Mondschein, Collin Kurre, Dag Wiese Schartum, Daniel Kraus, Danielle Ireland-Piper, Daphne Keller, Dariusz Kloza, David Johnson, David Post, David Vaile, Diane Rowland, Dietmar von der Pfordten, Erik Planken, Erik Sinander, Eva Maas, Felicity Gerry, Florence Guillaume, Fred Cate, Gail Kent, Giancarlo Frosio, Giovanni Sartor, Graham Greenleaf, Graham Smith, Gun Svantesson, Guy Berger, Hans Svantesson, Hans van Loon, Heidi Beate Bentzen, Hein

Dries-Ziekenheiner, Helena Lybæk, Hielke Hijmans, Ian Walden, Ioannis Revolidis, Jan Kerkhofs, Jan Oster, Jan Trzaskowski, Jane Reichel, Jay Forder, Jef Ausloos, Jesper Hagen, Jim Corkery, Jon Bing, Joost Blom, Joost Raeven, Julia Hörnle, Julia Powles, Kai-Uwe Schrogl, Kevin McGillivray, Lani Cossette, Léa Richard, Lee Bygrave, Liane Colonna, Lodewijk van Zwieten, Loren Holly, Lucy Cradduck, Madeline Taylor, Maja Brkan, Marcus Grahn, Marieke Koekkoek, Mark Lange, Mark Zoetekouw, Marketa Trimble, Marta Pertegás Sender, Matthew Rimmer, Michael Bogdan, Michael Geist, Michael Jensen, Michael Olsson, Michael Seng, Michael Stürner, Michal Czerniawski, Mikael Törmä, Mistale Taylor, Monique Mann, Orla Lynskey, Pablo Mendes De Leon, Patrik Lindskoug, Paul De Hert, Paul Fehlinger, Pedro de Miguel Asensio, Pedro Verdelho, Peggy Valcke, Peter Blume, Peter Wahlgren, Philip Smart, Philippe Van Linthout, Radim Polčák, Reid Mortensen, Richard Garnett, Rita Matulionyte, Roger Clarke, Rosa Riedl, Stefan Grundmann, Stephan Hobe, Steven Freeland, Steven Saxby, Tereza Kyselovska, Thomas Biermeyer, Thomas John, Thomas Schultz, Tjabbe Bos, Tobias Mahler, Tom Ginsburg, Torsten Bettinger, Toshiyuki Kono, Uldis Kinis, Ulf Maunsbach, Uta Kohl, Veronika Krizova, Vincenzo Zeno Zencovich, Vivek Krishnamurthy, Warren B Chik, William van Caenegem, Zdenek Kucera, Zsolt Balogh, and Ängla Eklund.

In particular, I want to thank Lee Bygrave for writing such a thoughtful, insightful and stimulating foreword, Catherine Karcher for again doing a wonderful job with the editing, and the brilliant (and patient) people I have worked with at Oxford University Press, not least Natalie Patey. I am also indebted to Alexander Seger, Cathrine Bauer-Bulst, and Bertrand de La Chapelle for their kind comments for the back cover.

During the four-year research project that resulted in this book, I gave forty-six keynotes, conference presentations, panel appearances, seminars, and guest lectures in fifteen different countries. My participation in each event taught me something new, and I thank all those who were willing to discuss my ideas during the course of them. During the project I also had the privilege of being a visiting scholar/professor at Queen Mary Centre for Commercial Law Studies University of London (UK), the Norwegian Research Center for Computers and Law at the University of Oslo (Norway), the Swedish Law & Informatics Research Institute at Stockholm University (Sweden), Faculty of Law, Masaryk University (Czech Republic), and Georg-August-Universität (Germany); I have benefited greatly from my time at those excellent institutions.

Obviously, the views (and any mistakes) presented are attributable solely to me.

Last, but not least, I thank my wonderful family; Bianca, Freja, and Felix. I promise not to write any more books for a (little) while.

Table of Contents

About the Book

The main ambition of this book is to discuss a set of proposals I developed during my four-year (2012–16) full-time *Australian Research Council Future Fellowship*[3] on the topic of Internet jurisdiction. While several of the proposals have already been presented in journal articles, they have never before been offered together as a multifaceted and (relatively) coherent framework for reform in the field of Internet jurisdiction. This book will do so: it expands on the proposals and, for the first time, explores the connections between these different proposals. Further, one important aspect of the book is to illustrate, in detail, why territoriality is a bad fit with the Internet and, indeed, our modern world more generally. Thus, while the book will draw and expand upon aspects of previously published works, the book goes well beyond merely reproducing previously published works. Nevertheless, it is only appropriate to here list the key works that I draw and expand upon in this book:

- *Private International Law and the Internet* (3rd edn, Kluwer Law International 2016)
- *Extraterritoriality in Data Privacy Law* (Ex Tuto 2013)
- 'Cyberborders through "Code": An All-or-Nothing Affair?', in Uta Kohl (ed), *The Net and the Nation State—Multidisciplinary Perspectives on Internet Governance* (CUP 2017) 110–24
- 'Enforcing Privacy Across Different Jurisdictions', in David Wright and Paul De Hert (eds), *Enforcing Privacy: Regulatory, Legal and Technological Approaches* (Springer 2016) 195–222
- 'Legal Theories of Private International Law: Overview and Practical Implications for Internet Regulation', in Patrik Lindskoug et al (eds), *Essays in Honour of Michael Bogdan* (Juristförlaget i Lund 2013) 539–55
- 'Online Cross-Border Defamation Disputes', in Dan Svantesson and Stanley Greenstein (eds), *Nordic Yearbook of Law and Informatics 2010–2012: Internationalisation of Law in the Digital Information Society* (Ex Tuto 2013) 195–215
- 'B2C Sales and Consumer Protection', in Radim Polčák (ed), *Introduction to ICT Law—Selected Issues* (Mazarykova univerzita, Brno 2007) 85–102

[3] Project no FT120100583.

- 'Article 4(1)(a) "Establishment of the Controller" in EU Data Privacy Law—Time to Rein In this Expanding Concept?' *International Data Privacy Law* (2016) 6/3 210
- 'Against "Against Data Exceptionalism"' 10(2) Masaryk UJ L & Tech (2016) 200
- (With Lodewijk van Zwieten) 'Law Enforcement Access to Evidence via Direct Contact with Cloud Providers—Identifying the Contours of a Solution' *Computer Law & Security Review* 32 (2016) 671
- 'Nostradamus Lite—Selected Speculations as to the Future of Internet Jurisdiction' (2016) 10(1) Masaryk UJL & Tech 47
- 'Jurisdiction in 3D—"Scope of (Remedial) Jurisdiction" as a Third Dimension of Jurisdiction' (2016) 12(1) J Priv Intl L 60–76
- 'A New Jurisprudential Framework for Jurisdiction: Beyond the Harvard Draft' (2015) 109 AJIL Unbound 69
- 'A Jurisprudential Justification for Extraterritoriality in (Private) International Law' (2015) 13(2) Santa Clara J Intl L 517–71
- 'Limitless Borderless Forgetfulness? Limiting the Geographical Reach of the "Right to be Forgotten"' (2015) 2(2) Oslo L Rev 116–38
- 'Extraterritoriality and Targeting in EU Data Privacy Law: The Weak Spot Undermining the Regulation' (2015) 5(4) Intl Data Priv L 226–34
- 'The Holy Trinity of Legal Fictions Undermining the Application of Law to the Global Internet' (2015) 23(3) Intl JL & Info Tech 219–34
- 'Delineating the Reach of Internet Intermediaries' Content Blocking—"ccTLD Blocking", "Strict Geo-location Blocking", or a "Country Lens Approach"?' (2014) 11(2) SCRIPT-ed 153–70
- 'Sovereignty in International Law—How the Internet (Maybe) Changed Everything, but not For Long' (2014) 8(1) Masaryk UJL & Tech 13755
- 'Between a Rock and a Hard Place—an International Law Perspective of the Difficult Position of Globally Active Internet Intermediaries' (2014) 30 Compu L & Sec Rev 348–56
- 'The Extraterritoriality of EU Data Privacy Law—Its Theoretical Justification and Its Practical Effect on U.S. Businesses' (2014) 50(1) Stan J Intl L 53–102
- 'A "Layered Approach" to the Extraterritoriality of Data Privacy Laws' (2013) 3(4) Intl Data Priv L 278–86
- 'What is "Law", if "the Law" is Not Something That "Is"? A Modest Contribution to a Major Question' (September 2013) 26(3) Ratio Juris 456–85
- 'Time for the Law to take Internet Geo-Location Technologies Seriously' (December 2012) 8(3) J Priv Intl L 473–87

- 'Celebrating 20 Years of WWW—a Reflection on the Concept of Jurisdiction' (2012) 6(1) Masaryk UJL & Tech 177–90
- 'Pammer and Hotel Alpenhof—ECJ Decision Creates Further Uncertainty About when E-Businesses "Direct Activities" to a Consumer's State under the Brussels I Regulation' (June 2011) 27(3) Computer L & Sec Rev 298–304

Finally I wish to stress that my aim has been to keep this book short and accessible. As a result, some, indeed many (if not all), of the topics covered deserve to be dealt with in greater detail than I have done here. To cater for this, I include references to additional reading on most topics.

Table of Cases

AUSTRALIA

Commonwealth

New South Wales

AUSTRIA

CANADA

British Columbia

EUROPE

FRANCE

INTERNATIONAL COURT OF JUSTICE

PERMANENT COURT OF ARBITRATION

PERMANENT COURT OF INTERNATIONAL JUSTICE

UNITED KINGDOM

UNITED STATES

Table of Legislation

Table of Treaties

Selected Abbreviations

ACCC	Australian Competition and Consumer Commission
CJEU	Court of Justice of the European Union
CJPTA	Court Jurisdiction and Proceedings Transfer Act (British Columbia, Canada)
CLR	Commonwealth Law Reports
CNIL	Commission Nationale de l'Informatique et des Libertés
DPA	Data Protection Authority
EC	European Council
ECHR	European Convention on Human Rights
ECJ	European Court of Justice
ICCPR	International Covenant on Civil and Political Rights
ICJ	International Court of Justice
IOC	International Olympic Committee
MICA	Ministry of Information, Communications and the Arts (Singapore)
MLAT	Mutual Legal Assistance Treaty
NSWLR	New South Wales Law Reports
NSWSC	New South Wales Supreme Court
PDPB	Personal Data Protection Bill (Singapore)
SCC	Supreme Court of Canada
WTO	World Trade Organization

1

Introduction

It is fair to say that the topic of 'Internet jurisdiction' is currently gaining an unprecedented level of attention. Indeed, at the moment, Internet jurisdiction is one of the most important, and most talked about, topics in Internet law and related fields. As we are seeing relatively little progress by way of real solutions, this will probably continue to be the case for some time, even with the fatigue that normally slows down the number of publications on a 'hot' topic after a while.

So what then is the problem, one may ask. This is arguably a poorly designed question since we are in fact facing several types of problem in the Internet jurisdiction arena. The closest we can come to an answer to such an overly broad question is probably something along the following lines: *the problem with Internet jurisdiction is the difficulty in finding an appropriate balance of the various interests at stake, and to express that balance in a sufficiently clear and precise manner so as to provide adequate guidance for those expected to act in accordance with the expressed balance.*

As would be obvious to anyone with legal training, or indeed anyone with a sufficient degree of so-called 'common sense', this problem statement could well be the articulation of just about any legal challenge imaginable. As if that was not enough, even this problem statement points to more than one issue. We can consequently already reach some interesting conclusions.

First of all, the Internet jurisdiction problems are not unique. Addressing these problems will involve interest balancing just as is the case for all legal problems. Second, just as for law more broadly, one constant source of problems is found in the difficulty in expressing, in words, the interest balancing for which we aim. Finally, here we have clear evidence of the initial claim that there is not one single Internet jurisdiction problem but rather a suite of them. Moreover, this latter observation must doubtless influence how we approach the area of Internet jurisdiction and how we work towards solutions.

One of the more obvious reasons for the suite of Internet jurisdiction problems stems from the very nature of the topic. Jurisdictional issues arise in every area of substantive law that has some cross-border dimension, and while several issues are relevant in relation to jurisdiction for a range of different substantive law areas, we must always exercise great care in applying solutions from one context to another, and we must always keep in mind the diversity of challenges we face. Just consider these scenarios:

> A law enforcement agency in Belgium has a warrant to search a suspect's home. There they find a computer with active access to the suspect's files stored on a cloud service in Japan. Can the law enforcement personnel access those files?
>
> A person in Australia posts defamatory content on Facebook. The content is read in twenty-five different countries, including in New Zealand. Where can the defamed person, who lives in New Zealand, take action? Against whom? And should Facebook block/remove the content even if it is lawful in all countries other than in New Zealand?
>
> An e-business in Brazil operates a website entirely in Portuguese. A consumer in Portugal orders a pair of orthopaedic shoes from the website, and in doing so must provide details about the foot condition she has as well as a delivery address. Has the Brazilian business directed its activities towards Portugal in such a manner that it can be sued in Europe, and can it be required to abide by the EU's data protection Regulation?
>
> Russian hackers attack key facilities and infrastructure, including banks, hospitals, and the power grid, in Estonia. What country has jurisdiction? What laws apply?
>
> A patient in the US is examined by a doctor in India. The doctor bases her diagnosis partly on a Skype examination of the patient and partly on data communicated by the patient's smart watch. The diagnosis is mistaken, resulting in permanent kidney damage to the patient. What country has jurisdiction? What laws apply? What prospect does the patient have of getting a remedy?

These hypothetical fact scenarios do not share many features whatsoever. Yet, they all fall within the scope of this book—they are all focused on jurisdictional issues in the Internet context. As Kuner notes:

> The difficulty of applying and enforcing any regulatory system [...] to the Internet rests on the fact that its operation involves a highly fragmented universe of actors, norms, procedures, processes, and institutions, including many non-state entities (such as private companies, non-governmental organisations, academic institutions, standards organisations, and others).[1]

This is, indeed, a useful expression of the challenges we face when seeking to apply law online. And Kuner's observation serves as a useful starting place

[1] Christopher Kuner, 'The Internet and the Global Reach of EU Law' (1 February 2017), forthcoming in the Collected Courses of the Academy of European Law (OUP) 7 <https://ssrn.com/abstract=2890930> accessed 2 April 2017.

if we were to map out the key problems that need to be addressed as far as Internet jurisdiction is concerned. Some key problems that ought to not be overlooked include the following:

1. Where jurisdictional rules are broad in scope, they risk capturing conduct with which there is an insufficient degree of contact to justify the jurisdictional claim. This may lead to jurisdiction being exercised over parties that lack adequate notice.
2. Where jurisdictional rules are narrow in scope, they risk leaving victims/plaintiffs without judicial redress.
3. Jurisdictional rules address complex questions, yet to be practically useful, they need to be simple and easy to understand and apply.
4. The online environment lends itself to, and online communities expect, a great deal of cross-border communication, some of which leads to cross-border disputes. Thus, the Internet jurisdiction rules must be able to cope with a high volume of disputes.
5. The mobility of data undermines the utility of several traditional jurisdictional anchor points.
6. Each country may be seen to have the right to dictate what is available online in that country.
7. No actor has actual control, or the right to exercise control, over the Internet as a whole.
8. Ensuring legal diversity online may necessitate placing borders online.

Others may opt to express or organise these concerns differently. And it may well be quite easy to add further problems to this list. But few would disagree that they are indeed problems for Internet jurisdiction.

Be that as it may, in any exercise aimed at mapping out the key problems that need to be addressed in relation to Internet jurisdiction, I would start with the most obvious issue—present thinking on jurisdiction is anchored in an unhelpful adherence to territoriality. Online it may be difficult to ascertain where information is stored, where events take place, and where people responsible for various activities are located. These are all important practical problems that must impact the way we approach matters of Internet jurisdiction. However, they need to be kept separate from related theoretical problems such as that it is often difficult to localise an event as occurring in a particular place in the online environment. This latter complication is, I would argue, an even more serious challenge to our current paradigm of jurisdictional thinking.

1.1 Our Current Territoriality-Focused Paradigm

Under our current paradigm, territoriality is the starting point for, and cornerstone of, all matters of jurisdiction. Amongst the principles used to assess jurisdictional claims it quite simply has no equals. Indeed, territoriality seems largely to be viewed as the true jurisprudential basis for jurisdictional claims—a state has the exclusive right to regulate all that occurs in its territory for the simple reason that it occurs in its territory.

Yet, as already hinted, there is a longstanding and clearly increasing recognition that strict territoriality is ill-equipped for today's modern society.[2] This is particularly so given that the society of today is characterised by constant, fluid, and substantial cross-border interaction, not least via the Internet. Even though our current international law rules on jurisdiction—including the territoriality principle—appear to be made out of rubber when one considers the degree to which they are stretched, detracted, and stepped on, they

[2] See eg Günther Handl, Joachim Zekoll, and Peer Zumbansen (eds), *Beyond Territoriality* (Martinus Nijhoff 2012); Cedric Ryngaert, *Jurisdiction in International Law* (2nd edn, OUP 2015); Alex Mills, 'Rethinking Jurisdiction in International Law' (2014) 84 Brit YB Intl L 187; Anne-Marie Slaughter, 'The Future of International Law Is Domestic (or, The European Way of Law)' (2006) 47 Harv Intl LJ 327; Paul Schiff Berman, *Global Legal Pluralism: A Jurisprudence of Law beyond Borders* (CUP 2014); Jan Oster, 'Rethinking *Shevill.* Conceptualising the EU Private International Law of Internet Torts Against Personality Rights' (2012) 26 Intl Rev L, Compu & Tech 113; Thomas Schultz, 'Carving up the Internet: Jurisdiction, Legal Orders, and the Private/Public International Law Interface' (2008) 19 Eur J Intl L; Daniel Cooper and Christopher Kuner, 'Data Protection Law and International Dispute Resolution' (2017) 382 Recueil des cours 9–174; Cedric Ryngaert and Mark Zoetekouw, 'The End of Territory? The Re-Emergence of Community as a Principle of Jurisdictional Order in the Internet Era' in Uta Kohl (ed), *The Net and the Nation State: Multidisciplinary Perspectives on Internet Governance* (CUP 2017) 185–201; Mary Keyes, *Jurisdiction in International Litigation* (The Federation Press 2005) 181; Benedetta Ubertazzi, *Exclusive Jurisdiction in Intellectual Property* (Mohr Siebeck 2012); Alexander Orakhelashvili, 'State Jurisdiction in International Law: Complexities of a Basic Concept' in Alexander Orakhelashvili (ed), *Research Handbook on Jurisdiction and Immunities in International Law* (Edward Elgar 2015) 9; Cecilia Magnusson Sjöberg, 'Rättens internationalisering i det digitala informationssamhället—Folkrätt och rättsinformatik' in Rebecca Stern and Inger Österdahl (eds), *Folkrätten i svensk rätt* (2012) and, somewhat more carefully, Hannah L Buxbaum, 'Territory, Territoriality, and the Resolution of Jurisdictional Conflict' (2009) 57 Am J Comp L 631, 674 noting how '[t]he use of territoriality as a mechanical standard used to link particular conduct with a particular country's law—rightly rejected—must be distinguished from the use of territoriality as an expression of a specific understanding about fairness and legitimacy in cross-border regulation', and Daniel Bethlehem observing that 'the edifice of contemporary international law—with its central pivots of territorial sovereignty, territorial jurisdiction, territorial regulation, territorial accountability—faces important conceptual and practical challenges' ('Comments on Hurrell: The End of Geography?' in James Crawford and Sarah Nouwen (eds), *Selected Proceedings of the European Society of International Law*, vol 3 (Hart 2012) 22).

nevertheless do not provide flexibility in the right way to accommodate modern society.

In the light of this, I will here join the growing chorus calling for a paradigm shift in relation to the crucially important concept of jurisdiction. And, as perhaps the most significant contribution of this book, I will provide a sketch of what a new paradigm might look like. In doing so, I adopt what best can be described as a deconstructive analysis and aim to connect the jurisdictional issues of concern with both wider jurisprudential concerns and with some specific practical problems.

The jurisprudential framework for jurisdiction that I propose as the new paradigm consists of three core principles. Those principles are distilled from existing public and private international law principles, as well as associated international law concepts such as comity and the 'duty of non-interference'.

The application of the core principles will be illustrated and an attempt will be made to (1) pre-empt some of the potential objections that may be raised against the proposal, (2) canvass how this framework relates to key existing jurisdictional principles, and (3) concisely anchor the proposed jurisprudential framework in the literature.

Further, by proving that the proposed jurisprudential framework for jurisdiction can be applied not only to public international law but also and equally well to private international law, we see something going beyond a convergence between public and private international law. In fact, we have identified the common core that *unites* public and private international law as far as jurisdiction is concerned.

1.2 The Need for Broad Reform

While the most important proposal is the far-reaching paradigm shift referred to above, it is by no means the only reform proposal presented in this book. Indeed, it is complemented by a series of other, more limited, reform proposals that each contributes a piece to the solution of the Internet jurisdiction puzzle. Some of those proposals build on, or indeed stem from, the jurisprudential framework for jurisdiction advanced, while others are more 'freestanding'.

To prepare ground for the discussion of my reform proposals, I will provide a brief, admittedly incomplete and necessarily superficial, introduction to the history of Internet jurisdiction in which I focus on a series of key events. The aim in doing so is to set the scene for the present time, as well as to highlight some approaches that have been tried (with or without success) in the past.

However, for it to be meaningful to discuss the history of Internet jurisdiction, we need to have a basic understanding of the jurisdictional landscape as it looked at the time the Internet began to be recognised as something giving rise to jurisdictional conundrums. Thus, the next chapter is dedicated to providing such an overview. In particular, I will have reason to discuss the concept of territoriality in detail and how the so-called 'Harvard Draft' from 1935 and the Permanent Court of International Justice's 1927 decision in the *Lotus* case[3] cemented its status as the crown jewel of jurisdictional principles. In this context I also address the related concept of extraterritorial jurisdiction. More specifically, it will be illustrated that the conventional distinction between jurisdictional claims that are 'extraterritorial' and those that are 'territorial' is associated with two problems: (1) it simply does not work, and (2) it tells us nothing of significance. The discussion will also bring attention to the fact that a strict application of territoriality is destructive and that State practice has already moved beyond territoriality and territorial sovereignty. Furthermore, examples are provided demonstrating that the concept of 'territoriality' is a component of a broader, dysfunctional, system within international law.

After the history portion, I will proceed to discuss what I have termed 'jurisdictional interoperability', which I see as the necessary overarching guiding principle for the mid-term perspective. I will also link the discussion into more fundamental questions of what law is, what principles ought to guide how we interpret jurisdictional law, and the roles jurisdictional law fulfils. In that context, I draw attention to the distinction between, on the one hand, what we may term 'bite jurisdiction' and, on the other hand, what may be referred to as 'bark jurisdiction'. I will then outline several reform proposals that I see as important pieces we can use to solve the Internet jurisdiction puzzle. I argue that we should recognise a concept of 'market sovereignty', based on what I refer to as 'market destroying measures'. I argue that we need to reconsider the classification we apply to jurisdictional claims, and I advance the concepts of 'investigative jurisdiction' and 'scope of (remedial) jurisdiction'. I also promote a 'layered approach' to jurisdiction and a 'doctrine of selective legal compliance'; and I draw attention to the impact of the use of geo-location technologies.

My reform proposals may come across as drastic, amounting to big steps away from the law as it stands today (*de lege lata*). However, I think we must be careful about how we think of *de lege lata* in the Internet context. In some situations it is obvious that our traditional laws from the offline environment

[3] SS 'Lotus' (*France v Turkey*) (1927) PCIJ Series A, No 10.

apply also online. But quite often the application of *de lege lata* online needs to be preceded by an assessment of whether the offline law may sensibly be applied online. We may, for example, need to ask whether it makes sense to focus on the location of contract formation online or whether the so-called 'postal acceptance rule'[4] applies to email communications. On most such occasions, we are likely to find a way to apply the existing law in a sensible manner to the online activity. But sometimes new principles are developed for the online environment; consider, for example, the Court of Justice of the European Union's (CJEU's) creation of the 'centre of interest test'.[5]

The key thing to remember is this: whether we conclude that *de lege lata* can be applied online or not, we are expanding on existing law; a finding that a certain existing legal norm applies in the same manner online as it applies offline expands that legal norm's application into a new environment in relation to which it was previously not known whether or not it applied. Thus, it is necessary to confront the misleading portrayal of law as merely being *applied* where it is in fact *expanded*.

1.3 Fundamental Assumptions

Finally, by way of introduction, this book is written in the light of eight assumptions—some of which have already been hinted at—that for transparency reasons I feel obliged to mention already, here in the introduction, as these assumptions no doubt (1) have influenced how I have approached the writing of this book, and (2) constitute the background to the arguments and proposals I present in the book.

The first of these assumptions is that it is both possible and fruitful to approach Internet jurisdiction issues as a distinct and homogeneous field. The first part of this—the assumption that it is *possible* to do so—rests on another assumption that needs to be acknowledged. I am of the opinion that, as far as jurisdiction is concerned, there is no strong reason to strictly separate between public and private international law. In fact, in the context of jurisdiction, it is not entirely clear where public international law starts and where private international law ends. As to the fruitfulness of approaching

[4] 'Where the circumstances are such that it must have been within the contemplation of the parties that, according to the ordinary usages of mankind, the post might be used as a means of communicating the acceptance of an offer, the acceptance is complete as soon as it is posted', *Henthorn v Fraser* [1892] 2 Ch 27, 33.

[5] Cases C-509/09 *eDate Advertising GmbH v X* and C-161/10 *Olivier Martinez and Robert Martinez v MGN Limited.*

Internet jurisdiction issues as a distinct and homogeneous field, I argue that, since many of the jurisdictional challenges that the Internet has brought to private international law are the same as those it has brought to public international law, each field can obviously learn from how the other has sought to tackle those challenges. This latter aspect of the assumption ought to be rather uncontroversial. Yet, I would go one step further. I have no doubt that the jurisdictional challenges that face the Internet would be much better, and more efficiently, addressed if we managed to break down the illusion of a dividing barrier between jurisdictional issues arising within private international law and those arising within public international law. Time will tell.

The second of the assumptions that underpin this book is that cross-border interaction is beneficial. Without necessarily endorsing all Fuller's propositions, there is little doubt that he points to something of the greatest significance to our information age when he, well before the Internet, stated:

> Communication is something more than a means of staying alive. It is a way of being alive [. . .] [I]f I were asked, then, to discern one central indisputable principle of what may be called substantive natural law—Natural Law with capital letters—I would find it in the injunction: Open up, maintain, and preserve the integrity of the channels of communication by which men convey to one another what they perceive, feel, and desire.[6]

This—the central role of communication—must, in my view, necessarily be borne in mind when we discuss jurisdiction, especially in the context of the Internet which, after all, is the greatest facilitator of communication the world has ever seen. In other words, I am admittedly and openly pro-communication across borders.

A third assumption to be noted is that the Internet serves us best where online interactions are governed by law; in other words, a lawless Internet is not as beneficial for humanity as is an Internet governed by law. A lawless Internet cannot create trust and it cannot provide security, and thus we need laws. Law-sceptics will no doubt reply that, even though laws apply online, the Internet is nevertheless characterised by distrust and lack of security. They would of course be right in pointing this out. However, it is important to discuss trust/distrust and security/lack of security as matters of degree rather than as binary opposites. I argue that we today have greater trust in the Internet and benefit from greater online security than we would in the total absence of laws applying online. Furthermore, without laws applying online, big-scale online commerce would be unthinkable given that there could be no trust in that contracts were upheld. And those who now feel inclined to

[6] Lon L Fuller, *The Morality of Law* (2nd edn, Yale University Press 1969) 186.

make the point that the Internet should not be a tool for commerce anyhow may do well to consider the extent to which e-commerce has benefited the people in the poorer parts of the world and worked as an opportunity equaliser. Furthermore, on an Internet where laws did not apply we would have to be prepared to see, and accept, the vilest, most despicable content such as child abuse materials flourish. I for one cannot see such content as acceptable in any format, and the Internet cannot be different in this regard. The simple fact is, despite the obvious enforcement challenges, we need the law online as well as offline.

The assumption as to the central role of law should not, however, be seen as an endorsement of law to the exclusion of other forms of regulatory forces. I acknowledge the role of 'soft law' (although I am not fond of the term as such[7]). I also acknowledge the central role that contracts play in Internet regulation.[8] In other words, it is 'law versus lawlessness' I am dealing with here, not 'law versus other forms of regulation'.

Fourth, when reading literature discussing the Internet's impact on law, one easily gets the impression that all was working smoothly until the Internet; that the international law system was a smooth machine whose operation was, somehow, disrupted by the Internet. We must always remember that this is not so. By taking this utopian starting point, we are misled into assuming that we need to tame the Internet so as to allow us to return to the utopian pre-Internet state of things. This shapes our methods and our thinking.

The reality is of course quite different. The international law system has never worked well. It was never a smooth machine, and there is no utopian state to which we may return. As I will highlight, the international legal system is, and has always been, dysfunctional. And while admittedly a strong claim, this is not even a particularly controversial claim in international law circles (or indeed, in legal theory circles). It then seems important that this is acknowledged in the discussion of Internet jurisdiction.

In fact, it seems to me that many scholars outside the field of international law view it as a particularly complicated area of law. But perhaps one reason they feel that they 'don't get it' is found in the fact that 'it' does not make sense. There are simply so many aspects of how international law approaches jurisdiction that are illogical and dysfunctional that the 'uninitiated' are prevented from understanding the system—the systems flaws in how international law approaches jurisdiction work as a 'barrier to entry'. And perhaps

[7] It seems to me there is little point in overextending the meaning of the term 'law'.

[8] See further, Lee A Bygrave, *Internet Governance by Contract* (OUP 2015).

the impact of this barrier is made even worse by the great deference international law scholars are taught—indeed commanded—to show towards the teachings of 'highly qualified publicists'.[9]

The fifth assumption relates to the central role played by Internet intermediaries. Without such intermediaries in the form of search engines, auctioning platforms, social media platforms, etc., we—as Internet users—would have a considerably less useful Internet. Put bluntly, Internet intermediaries need to be protected, not for their own benefit but for the benefit of the Internet users. While this claim may look rather controversial at first sight, I am not stating anything new here. To see that this is so, one need only consider, for example, the well-known §230 of the US *Communications Decency Act* of 1996 and Articles 12–15 of the E-Commerce Directive.[10] Both these instruments provide Internet intermediaries with protection against liability in certain circumstances. In addition to this, I argue that we need to protect Internet intermediaries by (1) where possible, avoiding States 'outsourcing' compliance assessments and enforcement to Internet intermediaries; (2) ensuring that the law provides as clear as possible guidance as to what is expected of the Internet intermediaries; and (3) minimising, or preferably eliminating, situations where Internet intermediaries risk being held to be violating one State's law by complying with another State's law.

The sixth assumption I wish to highlight is that law is manmade. The importance of this truism is that we are then free to question, deconstruct, reassess, and, indeed, reform it. We have the ability to revoke, amend, or alter the law as we see fit, and we should not be afraid to do so where change is called for.

My seventh assumption is that the ultimate goal of international law is to help ensure the survival of the human species. Thus, its sub-goals include matters such as ensuring peaceful coexistence, environmental protection, human rights, etc. The Internet can help by building links between different countries through cross-border communication, but can also be unhelpful when used as a new arena for international conflict. At any rate, given this assumption, jurisdictional rules must work towards this the ultimate goal of international law too.

The eighth and final assumption on which this book is built is perhaps somewhat more controversial than some of the others—in my view, our current focus on territoriality, driven as it is by national interest, is

[9] Statute of the International Court of Justice, Art 38(1).

[10] Directive (EC) 2000/31 of the European Parliament and Council, 8 June 2000, on Certain Legal Aspects of Information Society Services, in particular Electronic Commerce [2000] OJ L178/1, 369.

unsustainable. Much of the work in further chapters is directed at this very topic. But put in the fewest words, I am 'territoriality-agnostic', or perhaps even a 'territoriality-nihilist'.

Many, if not all, of the assumptions described probably fall into the category of 'common sense'. However, as suggested by Norwegian engineer Jens Olav Walaas Selvaag, if all else fails, try common sense![11]

Summarising the assumptions outlined, we may end up with the following declaration, a declaration that concludes this the first chapter and sets the scene for what is to come in the further text:

> *The Internet serves humanity best where it caters for cross-border interaction governed by law: law with an appropriate reach determined not merely by national territoriality but also by broader international interests that transcend the public/private international law divide.*

[11] Jens Olav Walaas Selvaag (1912–2002), author's translation of *'Hvis alt annet slår feil, prøv med sunn fornuft!'*

2

The Tyranny of Territoriality

It may sound overly dramatic, but I argue that, as far as jurisdictional thinking is concerned, our paradigm may be characterised as the tyranny of territoriality. We 'elected' territoriality as the guiding principle for jurisdiction in the belief that it would serve us; but instead, we find ourselves being ruled by territoriality in a tyrannical fashion and every move we make in the jurisdiction arena is restricted by the hard hand of territoriality. Thus, it is both timely and fitting to recall Edmund Burke's observation that: 'All tyranny needs to gain a foothold is for people of good conscience to remain silent.'[1] It is now time to break the silence.

The focus on territory runs like a *fil rouge* throughout contemporary private and public international law, and in some areas of law, territoriality is, if possible, even more pronounced than is generally the case. Looking at public international law, it has been noted that '[e]nforcement jurisdiction is, in international law, almost exclusively territorial.'[2] Yet there is nothing inevitable about territoriality, and we must recall that there was such a thing as a pre-territoriality period: 'From ancient times a person was subject to the law of his tribe. This law followed the person wherever he moved. However, over time, both the influence of and the connection to the tribe faded. The principle of personality weakened and the principle of territory grew stronger.'[3]

[1] <https://www.brainyquote.com/quotes/quotes/e/edmundburk136431.html>, accessed 2 April 2017.

[2] Alex Mills, 'Rethinking Jurisdiction in International Law' (2014) 84 Brit YB Intl L 195. See also Cedric Ryngaert, *Jurisdiction in International Law* (2nd edn, OUP 2015) 195. To overcome some of the obvious downsides of this, states often enter into agreements, such as Mutual Legal Assistance Treaties (see further, Anna-Maria Osula, 'Mutual Legal Assistance & Other Mechanisms for Accessing Extraterritorially Located Data' (2015) 9 Masaryk UJL & Tech 43, and Gail Kent, 'Sharing Investigation Specific Data with Law Enforcement–An International Approach' (Stanford Public Law Working Paper, 14 February 2014) <http://ssrn.com/abstract=2472413> or <http://dx.doi.org/10.2139/ssrn.2472413>, accessed 2 April 2017).

[3] Berte-Elen R Konow, 'Observations on Some Points of Contact between Private and Public International Law in Sweden' in Patrik Lindskoug et al (eds), *Essays in Honour of Michael Bogdan* (Juristförlaget 2013) 207.

Solving the Internet Jurisdiction Puzzle. First Edition. Dan Jerker B. Svantesson.

Indeed, in his thought-provoking and detailed discussion of the history of territorial jurisdiction, Ford notes, for example, that:

> the Thai state did not develop a concept of territorial jurisdiction until the latter part of the nineteenth century. Until then, Siam had none of the characteristics of jurisdictional authority. It did not control a contiguous territory defined by fixed and objective borders. Instead, it controlled a set of specific, non-adjacent places according to their proximity and usefulness to Bangkok; it controlled specific resources, trade routes or populations. It did not conceive of its authority in terms of territory. Instead, political authority operated by status hierarchy, with the elites in Bangkok at the top and various minor rulers occupying tiers in a dynastic pyramid. [. . .] Until at earliest the tenth or eleventh century and perhaps as late as the fifteenth century, European conceptions of political space were much like those of pre-modern Siam. Space was understood only in concrete terms or in relation to the plan of the divine. In pre-modern Europe, what appear to modern eyes to be territorial communities were in fact simply groups united by kinship, common interests and customs.[4]

It is not my ambition here to provide a detailed history of the concept of territoriality.[5] However, while the quotation makes clear that the territoriality principle was introduced at different times in different places to different degrees in different areas of law, I acknowledge that the history of the concept of territoriality is rather consistent from its inception. Put differently, the concept of territoriality we work with today is linked to its historical origins through a rather steady stream of endorsements by courts, legislators, policymakers, and scholars.[6] However, as I will return to in more detail, there are reasons why we must not exaggerate the importance of this.

While there thus can be no dispute about the strong current position held by the principle of territoriality, it is worth recalling that at the time of the birth of private international law, it was already recognised that, for trade to run as smoothly as possible, neither the principle of personality nor the principle of territoriality 'could be applicable alone or in their purest content'.[7] And in the words of Handl:

> [T]erritorial sovereignty could never convincingly be posited as an absolute principle. Instead, it was understood to require some adjustment in view of states' physical co-existence and associated transboundary spill-over effects, and the desirability of

[4] Richard T Ford, 'Law's Territory (A History of Jurisdiction)' (1998–99) 97 Mich L Rev 843, 868–72.

[5] See instead eg Ryngaert (n 2) 50–62.

[6] See eg Alex Mills noting how '[t]he idea of territoriality was expressed, for example, in the first two "maxims" of the Dutch eighteenth century private international law scholar Ulrich Huber' ('Rethinking Jurisdiction in International Law', 202).

[7] Berte-Elen R Konow (n 3) 208.

international trade and diplomatic intercourse. In short, 'extra-territoriality' was always an indispensable, complementary legal notion, the proverbial exception to the rule of territorial sovereignty's exclusiveness. By artfully tweaking the twin parameters of territoriality/extra-territoriality modern states for a long time managed to agree on mutually-acceptable limits to the exercise of what might be called transnational legal authority while re-validating the basic principle of territorial sovereignty.[8]

Furthermore, history also highlights that the law is familiar and comfortable with jurisdictional claims going beyond, or perhaps above, territoriality. The most obvious example is how the Church exercised a sphere of jurisdiction that was not anchored in territoriality.[9] However, I hasten to point out that I am by no means advocating a return to such a structure. Scholars seeking their inspiration in the role of jurisdictional claims of the Church do well to remember that Canon law never faced any enforcement difficulties whatsoever, given that offenders could expect to be condemned to spend eternity in a perceived 'hell'.

Looking at policy documents, and academic writings, the most commonly cited sources of our territoriality focus are the 1927 *Lotus* case,[10] and the 1935 'Harvard Draft Convention on Jurisdiction with Respect to Crime' (the 'Harvard Draft').[11] However, all too often these authorities are used in a rather too cavalier a fashion with little or no genuine understanding of the severe limitations of these 'key' authorities. Here, I will therefore first seek to dissect these authorities in some detail in an attempt to show them in their true light. I will then be able to proceed to consider more broadly, and bring attention to a range of aspects of, the role that territoriality plays in contemporary law.

2.1 The Influence of Colliding Steamships

The *Lotus* case,[12] decided by the then Permanent Court of International Justice in 1927, involved the collision between a French mail steamship (the *Lotus*) and a Turkish collier steamship (*Boz-Kourt*). The collision occurred just before midnight on the high seas on 2 August 1926. Following the collision, the Turkish steamer was cut in two and sank, with eight Turkish

[8] Günther Handl, 'Extra-territoriality and Transnational Legal Authority' in Günther Handl, Joachim Zekoll, and Peer Zumbansen (eds), *Beyond Territoriality* (Martinus Nijhoff 2012) 4 (footnote omitted).

[9] See eg PE Nygh, 'The Territorial Origin of English Private International Law' (1964) 2 U Tas L Rev 28, 30, and 32.

[10] SS 'Lotus' (*France v Turkey*) (1927) PCIJ Series A, No 10.

[11] 'Introductory Comment to the Harvard Draft Convention on Jurisdiction with Respect to Crime 1935' (1935) 29 Supp Am J Intl L 443.

[12] SS 'Lotus' (*France v Turkey*) (1927) PCIJ Series A, No 10.

nationals killed. Upon the French steamer's arrival in present-day Istanbul, Turkey instituted criminal proceedings in Turkey, in pursuance of Turkish law, against the officer of the watch on board the *Lotus* and against the captain of the *Boz-Kourt*. The resulting matter coming before the Court primarily related to whether Turkey's claim of jurisdiction was in conflict with the principles of international law.

2.1.1 The Majority Opinion

The majority judgment contains a discussion that, although lengthy, deserves to be reproduced in full:

International law governs relations between independent States. The rules of law binding upon States therefore emanate from their own free will as expressed in conventions or by usages generally accepted as expressing principles of law and established in order to regulate the relations between these co-existing independent communities or with a view to the achievement of common aims. Restrictions upon the independence of States cannot therefore be presumed.

Now the first and foremost restriction imposed by international law upon a State is that – failing the existence of a permissive rule to the contrary – it may not exercise its power in any form in the territory of another State. In this sense jurisdiction is certainly territorial; it cannot be exercised by a State outside its territory except by virtue of a permissive rule derived from international custom or from a convention.

It does not, however, follow that international law prohibits a State from exercising jurisdiction in its own territory, in respect of any case which relates to acts which have taken place abroad, and in which it cannot rely on some permissive rule of international law. Such a view would only be tenable if international law contained a general prohibition to States to extend the application of their laws and the jurisdiction of their courts to persons, property and acts outside their territory, and if, as an exception to this general prohibition, it allowed States to do so in certain specific cases. But this is certainly not the case under international law as it stands at present. Far from laying down a general prohibition to the effect that States may not extend the application of their laws and the jurisdiction of their courts to persons, property and acts outside their territory, it leaves them in this respect a wide measure of discretion, which is only limited in certain cases by prohibitive rules; as regards other cases, every State remains free to adopt the principles which it regards as best and most suitable.

This discretion left to States by international law explains the great variety of rules which they have been able to adopt without objections or complaints on the part of other States; it is in order to remedy the difficulties resulting from such variety that efforts have been made for many years past, both in Europe and America, to prepare conventions the effect of which would be precisely to limit the discretion at present left to States in this respect by international law, thus making good the existing lacunae in respect of jurisdiction or removing the conflicting jurisdictions arising from the diversity of the principles adopted by the various States.

In these circumstances, all that can be required of a State is that it should not overstep the limits which international law places upon its jurisdiction; within these limits, its title to exercise jurisdiction rests in its sovereignty.[13]

This tells us several things of great significance that may usefully be expressed through 'codification':

RULE 1

Due to the nature of international law and the nature of states, restrictions upon the independence of states cannot be presumed.

RULE 2

As a default rule, a state may not exercise its power in any form in the territory of another State.

RULE 3

Rule 2 may be departed from only where there is a permissive rule, derived from international custom or from a convention, to the contrary.

RULE 4

As a default rule, a state is free to exercise jurisdiction (in the sense of extending the application of their laws and the jurisdiction of their courts) in its own territory, in respect of any case which relates to persons, property and acts outside their territory (including to acts which have taken place abroad).

RULE 5

Jurisdiction under Rule 4 is restricted only where there is a prohibitive rule, derived from international custom or from a convention, to the contrary.

All this is important. Yet, there are several noteworthy limits to what is said here. First of all, it should be noted that leading scholars—such as Ryngaert and Mann—have questioned whether the *Lotus* decision remains good law.[14] Further, while the Court discusses the implications of possible permissive rules, it does not specifically discuss what such rules might look like. Indeed, the Court does not even suggest that there actually are any such permissive rules. Similarly, while the Court here discusses the implications of possible prohibitive rules, it does not specifically discuss what such rules might look like, nor even suggest that there actually are any such prohibitive rules on a

[13] ibid 19.

[14] Ryngaert (n 2) 34, and Francis Mann, 'The Doctrine of Jurisdiction in International Law' in Karl M Meesen (ed), *Extraterritorial Jurisdiction in Theory and Practice* (Kluwer Law International 1996) 64, 66.

general level. Indeed, discussing whether there is any prohibitive rule in the specific context of the *Lotus* case, the Court concludes that there is no evidence of any principle of international law precluding Turkey from acting in the way it had.[15]

In addition, while we so far have focused on what the Court said, it is also worth considering what it did not say. For example, the Court could have, but did not, make any reference to the concept of comity. Further, the Court could have, but did not, make any reference to the principle of non-intervention.

Moreover, the majority opinion in *Lotus* leaves one crucial question unanswered. How do one identify the location at which a state is exercising jurisdiction? Imagine, for example, that a Bulgarian law enforcement agency has a warrant to search a suspect's house. There they find a laptop computer that is turned on and is logged on to a Dropbox cloud storage account. If the law enforcement staff in such a situation downloads content stored on Dropbox servers in the US, where is it exercising jurisdiction? In Bulgaria? In the US? Somewhere else? If in Bulgaria, Rule 4 above suggests that the activity is perfectly legal under international law unless someone can point to a relevant prohibitive rule. In contrast, if it is concluded that the exercise of jurisdiction takes place in the US, the mentioned Rule 2 suggests that the activity is contrary to international law unless someone can point to a relevant permissive rule. In other words, the majority opinion in the *Lotus* case merely sets the parameters within which answers may be found for a scenario such as that just described. Thus, even if one focuses exclusively on the majority opinion in the *Lotus* case, it may be that more has been read into that decision than the holding required.

In contrast, while the decision does not illuminate anything as to how we should identify the location at which a state is exercising jurisdiction, it does tell us something important about how we identify the location of an offence:

> No argument has come to the knowledge of the Court from which it could be deduced that States recognize themselves to be under an obligation towards each other only to have regard to the place where the author of the offence happens to be at the time of the offence. On the contrary, it is certain that the courts of many countries, even of countries which have given their criminal legislation a strictly territorial character, interpret criminal law in the sense that offences, the authors of which at the moment of commission are in the territory of another State, are

[15] SS 'Lotus' (*France v Turkey*) (1927) PCIJ Series A, No 10, 31.

nevertheless to be regarded as having been committed in the national territory, if one of the constituent elements of the offence, and more especially its effects, have taken place there.[16]

And with equal clarity, and with even greater implications, the majority also emphasised that '[t]he territoriality of criminal law [...] is not an absolute principle of international law and by no means coincides with territorial sovereignty.'[17] The importance of this statement can scarcely be overstated; it is nothing short of a direct contradiction of the assumption of the *Lotus* case as an authority for strict territoriality. From all this, we may deduce another Rule to be added to those already canvassed:

RULE 6

Under international law, offences, the authors of which at the moment of commission are in the territory of another state, may nevertheless be regarded as having been committed in the national territory, if one of the constituent elements of the offence, and more especially its effects, have taken place there.

On this point, the Court makes two observations worthy of repetition here. First, it notes that the 'effect' is a particularly important factor in relation to offences as they are punishable due to their effect rather than due to the subjective intention of the delinquent.[18] The Court referred specifically to manslaughter—which of course was the offence of concern in the *Lotus* case—but it is easy to think of other actions such as defamation, data privacy violations, or even contractual breaches, in relation to which the effect is the relevant aspect, rather than the intention. At the same time, while the majority's reasoning on this may come across as sensible on an initial reading, it is in fact highly questionable given that it means that it would be more difficult for a court to claim jurisdiction over a murder than over manslaughter, since intention is an element of the former but not the latter.[19]

Second, the Court pointed to the connection between acts and effects in situations such as the *Lotus* case:

These two elements are, legally, entirely inseparable, so much so that their separation renders the offence non-existent. Neither the exclusive jurisdiction of either State, nor the limitations of the jurisdiction of each to the occurrences which took place on the respective ships would appear calculated to satisfy the requirements of justice and effectively to protect the interests of the two States. It is only natural that each

[16] SS 'Lotus' (*France v Turkey*) (1927) PCIJ Series A, No 10. [17] ibid 20.
[18] ibid 24.
[19] See further the dissenting opinion of Moore (SS 'Lotus' (*France v Turkey*) (1927) PCIJ Series A, No 10, 83).

should be able to exercise jurisdiction and to do so in respect of the incident as a whole. It is therefore a case of concurrent jurisdiction.[20]

From this, I wish to draw out another Rule even though many readers may find it both mundane and overly obvious:

RULE 7

International law allows concurrent jurisdiction.

The importance of this observation will become apparent in what follows.

Finally, as to the majority judgment in the *Lotus* case, I feel it appropriate to emphasise the concern that the Court expressed about requiring jurisdictional claims to be based on universally accepted rules. They noted that doing so may, in practice, in many cases result in paralysing the action of the courts.[21]

2.1.2 Dissenting Opinions

Those inclined to make absolute statements regarding jurisdiction based on the *Lotus* case frequently fail to mention that no less than half the members of the Court expressed dissenting opinions. Without pausing to consider the finer points of how international law evolves, we may perhaps, at the minimum, conclude that the high number of dissenting opinions shows that the majority view is not entirely uncontroversial. In my humble opinion, this must necessarily affect how we relate to the *Lotus* decision.

Of the dissenting opinions, Lord Finlay did not even agree with how the majority had interpreted the issue before the Court.[22] He also disagreed with other fundamental assumptions made by the majority. For example, he suggested that '[t]he jurisdiction over crimes committed on a ship at sea is not of a territorial nature at all.'[23]

In his dissenting opinion, Loder seemingly called into question the most fundamental aspect of the majority opinion arguing:

> The family of nations consists of a collection of different sovereign and independent States.
>
> The fundamental consequence of their independence and sovereignty is that no municipal law, in the particular case under consideration no criminal law, can apply or have binding effect outside the national territory.

[20] SS 'Lotus' (*France v Turkey*) (1927) PCIJ Series A, No 10, 31. [21] ibid 20.
[22] ibid 52. [23] ibid 53.

This fundamental truth, which is not a custom but the direct and inevitable consequence of its premise, is a logical principle of law, and is a postulate upon which the mutual independence of States rests.[24]

Looking at this statement carefully, it is possible to argue that it can be reconciled with the majority view. After all, as we noted, Rule 2 derived from the majority opinion proclaims that as a default rule, a state may not exercise its power in any form in the territory of another state, which is fully in line with Loder's mentioned statement, as well as with his observation that '[t]he criminal law of a State [...] cannot extend to offences committed by a foreigner in foreign territory, without infringing the sovereign rights of the foreign State concerned, since in that State the State enacting the law has no jurisdiction.'[25]

Loder also recognises that the default rule may be affected by permissive rules and acknowledges that: (1) 'The criminal law of a State may extend to crimes and offences committed abroad by its nationals, since such nationals are subject to the law of their own country'[26] (the nationality principle) and that (2) state practice has moved beyond 'the strict rule governing the jurisdiction over offences committed by foreigners abroad [...] in so far as they are directed against the State itself or against its security or credit'[27] (the protective principle). All this fits perfectly well with Rule 3—relating to permissive rules—derived from the majority opinion. Furthermore, as is discussed in more detail further, it seems Loder here is discussing legislative jurisdiction, while the majority dealt with the *Lotus* case as a matter of adjudicative jurisdiction. Perhaps this further assists attempts to reconcile the different judgments.

So, in what do we find the disagreement between the dissenting opinions and the majority? The answer to this question needs to be broken down into two parts. First, and more importantly, many of the dissenting opinions viewed the Turkish claim of jurisdiction as a violation of the territoriality principle that, by virtue of the legal fiction under which ships constitute floating extensions of the flag state's territory, would give jurisdiction to France.[28] Thus, in pointing to the territoriality principle, they had identified a principle of international law that they saw as restricting Turkey's jurisdictional claim.

Second, Loder, for example, had a radically different view from that of the majority about the localisation of the offence, a matter he describes as being of 'capital importance for the purposes of the decision of the dispute before the Court'.[29] In Loder's view:

[24] ibid 34–35. [25] ibid 35. [26] ibid 35. [27] ibid 35–36.
[28] See eg ibid 38 (Loder), 47 (Weiss), and 63 (Nyholm). [29] ibid 37.

> It is clear that the place where an offence has been committed is necessarily that where the guilty person is when he commits the act. The assumption that the place where the effect is produced is the place where the act was committed is in every case a legal fiction. It is, however justified where the act and its effect are indistinguishable, when there is a direct relation between them; for instance, a shot fired at a person on the other side of a frontier; a parcel containing an infernal machine intended to explode on being opened by the person to whom it is sent. The author of the crime intends in such cases to inflict injury at a place other than that where he himself is.[30]

Loder does not see the act and effect in the *Lotus* case as indistinguishable as did the majority, but there is agreement that occurrences exist where the act and effect are indistinguishable and that such cases give rise to concurrent jurisdiction. Further, Loder emphasises that intention is an important ingredient in the finding of the act and its effect being indistinguishable. This is, of course, completely opposite to how the majority emphasised that the 'effect' is a particularly important factor in relation to offences punishable due to their effect rather than due to the subjective intention of the delinquent.

2.1.3 Concluding Observations as to *Lotus*

Having dissected the *Lotus* decision, it seems clear beyond intelligent dispute that the *Lotus* judgment is not as solid a foundation for the strict application of the territoriality principle as many may have presumed; and this is so especially in the Internet environment. It is a judgment containing contradictions and areas of lack of clarity. And it is a judgment made for a different era and for a specific and different context.

Now imagine that we were not bound by any existing law and were trying to design, from scratch, rules for jurisdictional claims in relation to Internet activities such as law enforcement cross-border access to data. What types of earlier cases could then provide useful guidance? It seems unlikely that we would turn to cases concerning colliding steamships as our primary source

[30] ibid 37. See also eg Weiss's opinion: 'In fixing the place where an offence has occurred, it is not to the place where the offence, often contrary to any reasonable anticipation, produces its injurious effects upon persons or things, that attention must be directed, but solely to the place where the punishable act has been committed and where the person responsible for that act was at the time when it was committed; it is there that the offence has really taken place' (ibid 47). But see also Moore's opinion: 'But it appears to be now universally admitted that, where a crime is committed in the territorial jurisdiction of one State as the direct result of the act of a person at the time corporeally present in another State, international law, by reason of the principle of constructive presence of the offender at the place where his act took effect, does not forbid the prosecution of the offender by the former State, should he come within its territorial jurisdiction' (ibid 73).

of inspiration. Put differently, cases concerning colliding steamships involve quite different considerations to those involved in the Internet jurisdiction context.

Here we may pause to consider an often overlooked, but highly relevant, observation made by the majority in the *Lotus* case. Discussing the necessity of ascertaining whether or not under international law there is a principle that would have prohibited Turkey from prosecuting the officer of the watch on board the *Lotus*, they noted: 'This must be ascertained by examining precedents offering a *close analogy* to the case under consideration; for it is only from precedents of this nature that the existence of a general principle applicable to the particular case may appear.'[31] Furthermore, the majority judgment repeatedly stresses that its 'examination is strictly confined to the specific situation in the present case, for it is only in regard to this situation that its decision is asked for.'[32]

Colliding steamships do not offer any *close analogy* to Internet jurisdiction cases. In the light of this, the *Lotus* case cannot—on the view of the majority in the *Lotus* case—constitute a precedent establishing the existence of any general principles of international law applicable to Internet jurisdiction cases.

To this may be added that Mills has argued that the *Lotus* case is an anomaly and that it 'can best be characterised as the "high water mark" of a briefly dominant but still highly influential theoretical approach to international law, generally known as international legal positivism'.[33]

Given all this, it is both surprising and disappointing to see how often the *Lotus* case is relied upon as an exclusive authority for broad-brushed statements as to jurisdiction, including in relation to Internet jurisdiction specifically.[34] After all, as is discussed below, for more than twenty years it has been recognised that the territoriality principle is a poor fit for the largely

[31] ibid 21 (emphasis added). [32] ibid 22.

[33] Mills, 'Rethinking Jurisdiction in International Law', 191.

[34] eg a Report of the Council of Europe's Transborder Group adopted by the T-CY on 6 December 2012 notes: 'If the data is stored in a foreign jurisdiction, LEA would need to resort to international cooperation. The basic rule of international law regarding the exercise of coercive powers is the principle of territorial sovereignty. No State may enforce its jurisdiction within the territory of another sovereign State. [footnote reading: 'Cf. Case of the SS "Lotus" (France v Turkey), PCIJ Series A, No 10'] As a consequence, international cooperation is dependent on international treaties—such as the Budapest Convention on Cybercrime—or bilateral agreements between the States concerned (Cybercrime Convention Committee (T-CY) Ad-hoc Sub-group on Jurisdiction and Transborder Access to Data' <http://www.coe.int/t/dghl/cooperation/economic-crime/Source/Cybercrime/TCY/TCY2012/TCY_2012_3_transborder_rep_V31public_7Dec12.pdf> accessed 2 April 2017).

The phrase 'enforce its jurisdiction' is vague at best, and there are many assumptions baked into this paragraph.

border-disregarding Internet, and as just noted, the facts of the *Lotus* case are in no way analogous to the Internet jurisdiction scenarios we seek to address at the present, such as law enforcement cross-border access to cloud data, requests for cross-border content removal, or Internet defamation disputes.

General legal method allows us to decide different cases differently. What is then the rationale for insisting that the *Lotus* decision compels us to view, for example, law enforcement cross-border access to data as strictly regulated by the territoriality principle, especially when other areas of law do not follow that principle in an equally strict manner?

2.2 The Influence of the 'Harvard Draft'

In 1935 a major step was taken in international law: the 'Harvard Research Draft Convention on Jurisdiction with Respect to Crime' was published. The influence of the 'Harvard Draft' has been nothing less than phenomenal and must surely have exceeded the drafters' wildest ambitions. It is fair to say that the structure put forward in the 'Harvard Draft' has represented public international law's approach to jurisdiction ever since.

Essentially, the lasting legacy of the 'Harvard Draft' is that it identified and articulated a set of five grounds for jurisdiction to varying degrees recognised under international law at the point in time it was drafted:

> These five general principles are: first, the territorial principle, determining jurisdiction by reference to the place where the offence is committed; second, the nationality principle, determining jurisdiction by reference to the nationality or national character of the person committing the offence; third, the protective principle, determining jurisdiction by reference to the national interest injured by the offence; fourth, the universality principle, determining jurisdiction by reference to the custody of the person committing the offence; and fifth, the passive personality principle, determining jurisdiction by reference to the nationality or national character of the person injured by the offence.[35]

But not all these principles were held to be of equal strength. Indeed, the legitimacy of all principles was not universally recognised:

> Of these five principles, the first [the territorial principle] is everywhere regarded as of primary importance and of fundamental character. The second [the nationality principle] is universally accepted, though there are striking differences in the extent to which it is used in the different national systems. The third [the protective

[35] 'Introductory Comment to the Harvard Draft Convention on Jurisdiction with Respect to Crime 1935' (1935) 29 Supp AJIL 443, 445.

principle] is claimed by most States, regarded with misgivings in a few, and generally ranked as the basis of an auxiliary competence. The fourth [the universality principle] is widely though by no means universally accepted as the basis of an auxiliary competence, except for the offence of piracy, with respect to which it is the generally recognized principle of jurisdiction. The fifth [the passive personality principle], asserted in some forms by a considerable number of States and contested by others, is admittedly auxiliary in character and is probably not essential for any State if the ends served are adequately provided for on other principles.[36]

While it is difficult to tell whether the above reference to the territorial principle was to be read to include so-called objective territoriality, particularly considering the *Lotus* case, it is hardly probable that the 'Harvard Draft' should be read to suggest that objective territoriality was 'everywhere regarded as of primary importance and of fundamental character'. Thus, I argue that there are in fact six, rather than five, such principles with the mentioned territoriality principle actually better split into two—subjective territoriality and objective territoriality.

At any rate, in a sense the 'Harvard Draft' thus provides us with a hierarchical structure of the jurisdictional principles that it identifies. However, and this is important, it is a hierarchy or ranking of which principles are more widely recognised than others; it is neither a hierarchy or ranking of which principles have a stronger claim to legitimacy nor of which principles trump the others where they clash.

Discussing the influence of the 'Harvard Draft', it should also be noted how it 'has been instrumental in the permissive principles approach becoming the main framework of reference for assessing the legality of jurisdictional assertions';[37] that is, the 'Harvard Draft' is seen to merely outline grounds upon which jurisdiction *may* be based, without for that sake 'blacklisting' all other grounds for jurisdiction.[38] However, the drafters of the 'Harvard Draft' may not actually have intended such a distinction as they specifically noted that:

[o]ne avenue of approach emphasizes the idea of capacity to act in the exercise of competence, the other the idea of limitations upon capacity. Both ideas are implicit in the concept of competence and in the term which is used in this Convention to describe it. It has seemed appropriate, therefore, without further refinement, to use the term 'jurisdiction' to denote the competence of States which it is the object of this Convention to determine.[39]

[36] ibid. [37] Ryngaert (n 2) 35.

[38] Consider also the mentioned Permanent Court of International Justice's conclusion in the *Lotus* case (SS 'Lotus' (*France v Turkey*) 18–19).

[39] 'Introductory Comment to the Harvard Draft Convention on Jurisdiction with Respect to Crime 1935' (1935) 29 Supp AJIL 443, 468.

2.2.1 What the 'Harvard Draft' Set Out to Do

We can learn more about the sources relied upon for the 'Harvard Draft' and, even more importantly, the limited scope of the 'Harvard Draft', through this quote:

> An analysis of modern national codes of penal law and penal procedure, checked against the conclusions of reliable writers and the resolutions of international conferences or learned societies, and supplemented by some exploration of the jurisprudence of national courts, discloses five general principles on which a *more or less* extensive *penal* jurisdiction is claimed by States *at the present time*.[40]

In other words, the 'Harvard Draft' was intended to be limited thus:

1. It summarises and restates the legal practices at the time in the form of five general principles on which jurisdiction is claimed by states as of 1935;
2. It considers jurisdiction only in penal matters; and
3. The five principles are accepted at varying degrees.

To this may be added that it is generally regarded that the 'Harvard Draft' principles govern only prescriptive (or legislative) jurisdiction.

Furthermore, despite the almost divine status the 'Harvard Draft' principles enjoy today, they did not arrive on stone tablets from some Heaven; there is nothing inherently holy about these principles. In fact, as is appropriately acknowledged in the 'Introductory Comment to the Harvard Draft', the grounds for jurisdiction were reached by a series of compromises and simplifications.[41]

2.2.2 What the 'Harvard Draft' Actually Did

Despite the limited aim of the 'Harvard Draft', the jurisdictional principles it articulated have become something much larger than intended or, at least, than anticipated. The 'Harvard Draft' principles, with the possible addition of the so-called 'effects doctrine' (discussed further below), have been accepted as a definitive statement of when states may exercise prescriptive, and often also adjudicative, jurisdiction under international law. Indeed, in what Mills describes as 'a fairly ritualised account of the standard "heads" of jurisdiction',[42] the 'Harvard Draft' grounds for jurisdiction are frequently portrayed as applicable also to enforcement jurisdiction.

[40] ibid 445 (emphasis added). [41] See eg ibid 475, 487, and 497.
[42] Mills, 'Rethinking Jurisdiction in International Law', 188.

To mention but one example, in outlining the current jurisdictional landscape, the first version—that of 2013—of the influential *Tallinn Manual on the International Law Applicable to Cyber Warfare*[43] relied exclusively, though not expressly, on the principles advanced in the 'Harvard Draft'. Importantly, having noted that '[t]he term "jurisdiction" encompasses the authority to prescribe, enforce, and adjudicate',[44] the *Tallinn Manual* proceeded to discuss the 'Harvard Draft' grounds for jurisdiction without distinguishing between prescriptive jurisdiction, adjudicative jurisdiction, and enforcement jurisdiction—it treated the 'Harvard Draft' grounds for jurisdiction as applicable to all types of jurisdictional claims.[45]

The most obvious reason the 'Harvard Draft' has had such an enormous influence is of course found in the fact that it represents a sterling research effort performed by top scholars, resulting in a convenient summary of a complex topic and published in a prestigious outlet at the right moment in time. Nothing either could, or should, detract from that.

It is, of course, also the case that the central position of territoriality sits well with dominant legal and political state-focused theories in general. Indeed, it has been noted that 'territory, space, boundaries, and borders are inescapable companions of every human being, both regarding his physical and intellectual existence'.[46] While this claim may have less relevance in today's online environment, I need not quarrel with it here. Moreover, I think Grahn brings attention to something highly relevant when he—in the context of discussing how states, in a military sense, are fixated on their ability to assert dominance over their respective territories—points to underlying biological foundations of the territoriality thinking.[47]

At any rate, at the time the 'Harvard Draft' principles were articulated, the territoriality principle was much more easily applied than it is today. Thus, part of the structure presented in the 'Harvard Draft' was always bound to be uncontroversial at the time it was presented.

However, given that some of the jurisdictional principles presented in the 'Harvard Draft' remain controversial today,[48] it is perhaps surprising that the

[43] Michael N Schmitt (ed), *Tallinn Manual on the International Law Applicable to Cyber Warfare* (CUP 2013) 8–21.

[44] ibid 18.

[45] ibid 8–21. This issue has been addressed in the second version published in 2017 (Michael N Schmitt (gen ed), *Tallinn Manual 2.0 on the International Law Applicable to Cyber Operations* (CUP 2017).

[46] Daniel-Erasmus Khan, 'Territory and Boundaries' in Bardo Fassbender and Anne Peters (eds), *The Oxford Handbook of The History of International Law* (OUP 2012) 226.

[47] Jan-Olof Grahn, *Om Svensk Underrättelsetjänst* (Medströms Bokförlag 2016) 293.

[48] I say controversial because in reading the literature on the grounds for jurisdiction one cannot avoid standard statements outlining which of the grounds are more controversial than the

five-principle structure as such has not been subjected to any real degree of opposition.

As already hinted at above, one reason for this may perhaps be found in how scholars in the international law field view the writings of high-level commentators. Through Article 38 of the Statute of the International Court of Justice, scholars in the field of international law are taught deference to 'the teachings of the most highly qualified publicists of the various nations'.[49] While such teachings are merely a subsidiary means for the determination of rules of law, I suspect that Article 38 of the Statute of the International Court of Justice creates a different attitude—a different mentality—to the writings of high-level commentators than what is typically seen in other fields of law. This may well have helped minimise challenges to the 'Harvard Draft' principles.

However, I suspect that there is also another important, more practical, and less flattering, reason for the 'Harvard Draft's' longstanding influence: it is quite simply a textbook writer's dream come true. Issues of jurisdiction are always complex, and public international law is no exception. Thus, at least for basic treatises on international law, it has served commentators well to pretend that the 'Harvard Draft' provides a solid and comprehensive framework for determining the circumstances under which a state may claim jurisdiction under public international law, even though in many ways it is nothing but a surrogate for a comprehensive and internally logical framework. And through a mantra-like repetition of these principles in virtually every text and document on international law, the 'Harvard Draft' principles have been

others; and invariably, the objective territoriality principle, the passive personality principle, and the effects principle are said to be amongst the more controversial grounds for jurisdiction. See eg Danielle Ireland-Piper, 'Extraterritorial Criminal Jurisdiction: Does the Long Arm of the Law Undermine the Rule of Law?' (2012) 13 Melb J Intl L 1, stating eg: 'While subjective territoriality is rarely controversial, objective territoriality is more complicated' (10) and 'The existence and use of the passive nationality principle is particularly controversial' (13). I argue, however, that such generalised statements are pointless and detract from what is important. It must be remembered that the controversial effects doctrine may well be applied in an uncontroversial manner in the right circumstances, while jurisdiction based on the less controversial territoriality principle may be highly controversial in some circumstances—we must avoid classifying jurisdictional grounds as being either controversial or uncontroversial simply by reference to which of the jurisdictional principles they fit into.

[49] Statute of the International Court of Justice, Art 38(1). See further Hugh Thirlway, *The Sources of International Law* (OUP 2014) 126–28. One may of course ponder how one identifies these 'highly qualified publicists' and how one distinguishes such remarkable people from the more ordinary 'plain vanilla' publicists. In fact, making such a distinction is clearly wrong. Individual works of publicists must necessarily be judged by their respective standards and merits, not by reference to who wrote them; good legal reasoning deserves to be considered whether the author is a well-recognised international lawyer or a first-year law student.

transformed into something they should not have become. Students of international law will no doubt have appreciated this at exam time. However, on the whole it has done a disservice to the advancement of jurisdictional issues under international law, at least in recent years.

In the light of this, it is at least arguable that our current focus on territoriality as a foundation for jurisdiction is actually more of a mindset than based on solid international law principles. The time has come to put an end to this apparent sleepwalking acceptance of territoriality based on the 1927 *Lotus* case and the 1935 'Harvard Draft'.

2.3 The Problems with Territoriality

In the light of the above, some of the problems with the 'Harvard Draft' do not actually stem from the text as such; rather they result from how we use it and what we have allowed it to become. We have allowed the 'Harvard Draft' and the *Lotus* case to become cementations of 1920s–30s thinking about the world. Outdated principles—principles of a different world—are governing and, to great detriment, restricting our thinking today.[50] As an unsurprising consequence, the principles found in the 'Harvard Draft' are no longer part of any solution; they have become a part of the problem. Nowhere is this truth more obvious than in the context of jurisdictional claims in relation to the Internet. Indeed, our traditional focus on time[51] and space[52] is fundamentally

[50] For an interesting attack on territoriality based on the concept of 'community', see Paul Schiff Berman, *Global Legal Pluralism: A Jurisprudence of Law beyond Borders* (CUP 2014).

[51] While our focus on 'time' is less affected than our focus on 'space', examples can be found of how modern technology, such as the Internet, undermines also the time-focus to a strong degree. Consider, for example, the rich legal discussion of the proper formation of distance contracts, dating back to Roman times. A variety of legal solutions have appeared, such as the 'postal acceptance rule' (see eg *Henthorn v Fraser* [1892] 2 Ch 27, 33), to deal with the time difference characteristic of distance contracts in former times. Today, however, distance contracts are not necessarily associated with lengthy time delays, and much of the focus on time is of lesser significance now than it was historically.

[52] An interesting example of how our concept of 'space' is being altered beyond recognition is found in a discussion accompanying Act No 181 of 23 July 2014 On Cyber Security and Change of Related Acts (Act on Cyber Security) (Czech Republic): 'The meaning and aim of the law is protection of *the Czech cyberspace* to provide entities under the jurisdiction of the Czech Republic adequate tools and standards for cybernetic security of their information systems and electronic communication and undisturbed exercise of their right for informational self determination. The material authority dealing with legal relations to information and communication infrastructure is corresponding to this aim. The division of material authority is determined by the fact that *the cyberspace of the Czech Republic* can be divided into critical (with high significance for functioning of the state) and the remaining part (all other services of electronic communications and services of information society). The intent of the law incorporates both networks and

undermined by the nature of the Internet. Particularly the application of the territoriality principles—the central principle for prescriptive, adjudicative, and enforcement jurisdiction, and described in the 'Harvard Draft' as 'everywhere regarded as of primary importance and of fundamental character'[53]—to the Internet has been problematic.

So far, we have largely turned a blind eye to these matters. Now the time has come to turn our minds to them instead, and in the following sections, I will discuss some problems associated with our current focus on territoriality, as well as some factors that clearly undermine the legitimacy of such a focus.

2.3.1 The Territoriality Principle Cannot Alone Be the Jurisprudential Core of Jurisdiction

The territoriality principle cannot alone account for jurisdiction. If it could, all the other 'Harvard Draft' principles would be redundant. But as the 'Harvard Draft' project showed us, states have found it necessary to supplement the territoriality principle.

All this is uncontroversial, and pointing it out here may seem somewhat gratuitous. However, this fact alone—the territoriality principle's inability to, on its own, account for jurisdiction—is fatal for its perceived position as the jurisprudential core of jurisdiction.

Here I want to pause to make a few observations about the 'effects doctrine' (ie jurisdiction based on the fact that conduct outside the state has effects within the state)[54] alluded to already. After all, its development is

services of electronic communications as well as information systems comprising together critical information infrastructure as parts of critical infrastructure of *the Czech cyberspace*' (*National Security Authority: Legislative Intent of the Law on Cybernetic Security* (Czech Republic) <https://www.govcert.cz/download/legislativa/container-nodeid-660/legislative-intent-law-on-csen.pdf> 40 (emphasis added), accessed 10 April 2017). This statement is highly interesting for at least two reasons. First, and most obviously, the reference to the fact that the intent of the law incorporates both networks and services of electronic communications as well as information systems comprising together critical information infrastructure highlights a merger of the real (ie the infrastructure) and the virtual (eg data). Second, the repeated reference to a 'Czech cyberspace' or 'the cyberspace of the Czech Republic' shows a notion of territorial space in the virtual space of Cyberspace. However, occurring in virtual space, such a territoriality cannot be grounded in any principle resembling the traditional territoriality principle. Rather, it must be grounded in something else, and I will suggest that that something else takes the form of three core principles focusing on the connection and interest in the matter, as well as a balancing with other interests.

[53] 'Introductory Comment to the Harvard Draft Convention on Jurisdiction with Respect to Crime 1935' (1935) 29 Supp AJIL 443, 445.

[54] David J Gerber, 'Beyond Balancing: International Law Restrains on the Reach of National Laws' (1984) 10 Yale J Intl L 185, 190.

perhaps a particularly apt demonstration of the inadequacy of strict territoriality thinking.

Particularly in the field of competition law, or as it is also called, antitrust law, there are examples of state practice embracing the effects test for jurisdictional claims going beyond the relevant state's territorial borders.[55] However, despite a relatively widespread adoption,[56] some will argue that this practice is by no means widely spread, or accepted, enough to warrant the conclusion that it is an established jurisdictional ground under customary international law. Indeed, the effects doctrine is often seen as a relative newcomer amongst the jurisdictional principles. For example, Senz and Charlesworth claim that the effects doctrine 'was first introduced in *United States v Aluminum Co. of America ("Alcoa")* [148 F 2d 416 (2nd Cir 1945)]'.[57] The accuracy of this claim hinges on the exact definition of the doctrine being 'introduced'.

An inquiry into the roots of the effects doctrine shows that it has a long history indeed, arguably traceable at least to the era of Edward III (1312–77).[58] At any rate, over 100 years ago, in *Strassheim v Daily*,[59] the Supreme Court of the United States took the view that '[a]cts done outside a jurisdiction, but intended to produce and producing detrimental effects within it, justify a State in punishing the cause of the harm as if he had been present at the effect, if the State should succeed in getting him within its power.'[60] A clearer articulation of the effects doctrine—be as it may on a domestic level—is hardly necessary. Thus, the effects doctrine predates the 'Harvard Draft'.

[55] eg the US case, *Hartford Fire Ins Co v California* 509 US 764 (1993) and the European Case 89/85 *In re Wood Pulp Cartel* (1988) ECR 5193. See PJ McConnaughay, 'Reviving the "Public Law Taboo" in International Conflict of Laws' (1999) 35 Stan J Intl L and WS Dodge, 'Extraterritoriality and Conflict-of-Laws Theory: An Argument for Judicial Unilateralism' (1998) 39(1) Harv Intl LJ 101.

[56] The effects principle has been adopted in eg Germany, France, Switzerland, Denmark, Sweden, and the European Economic Community, as well as the US: Gerber (n 54) 201–02.

[57] Deborah Senz and Hilary Charlesworth, 'Building Blocks: Australia's Response to Foreign Extraterritorial Legislation' (2001) 2(1) Melb J Intl L 69, 81.

[58] See eg Gray J's interesting discussion in *Commonwealth v Macloon* [1869] 101 Mass, 1, which includes the following: 'In the beginning of the reign of Edward III., according to Chief Justice Scrope, if a man died in one county of a wound received in another, the murderer might be indicted and arraigned in the county where the death happened, "and yet the cause of his death began in the other county." Fitz. Ab. Corone, 373.'

[59] *Strassheim v Daily* (1911) 221 US 280.

[60] ibid 285. In support of this statement Justice Holmes pointed to these authorities: *Commonwealth v Smith* (1865) 11 Allen, 243, 256, 259; *Simpson v State* [1911] 92 Georgia, 41; *American Banana Co v United Fruit Co* [1909] 213 US 347, 356; *Commonwealth v Macloon* [1869] 101 Mass, 1, 6, 18.

At least as far as some torts, such as defamation, are concerned, the public international law distinction between the effects test and objective rules (particularly objective territoriality) is overly simplistic and misguided.[61] Presume that two different people have been defamed in a WWW publication. Person A initiates proceedings in a court of his state of domicile, while person B initiates proceedings in a court of the state in which she has her main reputation (which is not the same as her home state). Both courts accept jurisdiction on the basis that the damages occurred in the forum state. Which court in this situation relies upon the effects doctrine, and which court relies upon an objective rule? Depending on the substantive defamation law of each state (ie how the damaging act is defined), both can claim to have based their jurisdiction upon the 'accepted' objective rule, and perhaps even on the subjective rules.[62] In contrast to the classical example of a gunshot being fired across a border, it is not immediately clear what constitutes the harm (ie the objective outcome) in a cross-border defamation case.

Indeed, some scholars take the view that '[the effects doctrine] is also superfluous in the sense that it is hard to imagine a state having a legitimate jurisdictional interest in a conduct that would not otherwise be covered by the protective principle or another less contentious basis of extraterritorial jurisdiction.'[63]

This may be somewhat of an oversimplification, however, and my point is not that the existing overlap makes the effects doctrine superfluous. Imagine, for example, that the relevant arm in State A of a multinational business suffers harm due to data being lost because of a hacker attack on a cloud service provider based in States X, Y, and Z. Does State A have jurisdiction? To rely on the passive personality principle does not seem possible; after all, the business cannot be seen to be a national of State A merely by having a branch in State A. Similarly, the objective territoriality principle does not provide a ground for a jurisdictional claim by State A as the hacking crime is not consummated within State A's territory.[64] The effects doctrine would, however, provide such a base for jurisdiction as it is beyond intelligent dispute that a

[61] Those who have recognised the extremely strong connection between the effects doctrine and the objective rules have been accused of 'confusing' the two. See eg Gerber (n 54) 191.

[62] As pointed out by Gerber, the objective territoriality principle 'establishes the jurisdiction of the state to prosecute and punish for crimes commenced without the state but consummated within its territory': ibid 195, referring to 'Introductory Comment to the Harvard Draft Convention on Jurisdiction with Respect to Crime 1935' (1935) 29 Supp Am J Intl L 443, 435.

[63] Danielle Ireland-Piper (n 48) (2012) 13 Melb J Intl L 1, 17.

[64] As pointed out by Gerber, the objective territoriality principle 'establishes the jurisdiction of the state to prosecute and punish for crimes commenced without the state but consummated within its territory': Gerber (n 54) 194–95 referring to the 'Harvard Draft'.

substantial effect was felt in State A. Indeed, such an effect, in State A, may well have been foreseeable, or even the intended consequence of the hacking activity. I have no doubt similar examples can be found that predate cloud computing and, indeed, that predate Internet technology. However, maybe it can be said that Internet technology, and not least cloud computing, makes a move from a focus on nationality and territoriality to a focus on effect important.

In other words, while it has a long history, there are clear indications that the necessity of an effects-focus is increasing with globalisation, and it does so at the inevitable expense of the importance of strict territoriality.

2.3.2 A Strict Application of Territoriality is Destructive

As noted by Moore in his dissenting opinion in the *Lotus* case, 'In England it was once held that where a blow was struck in one county and death ensued in another county, the criminal could not be tried in either.'[65] This is but one illustration of the less than desirable implications that may flow from a strict application of the territoriality principle.

Thus, despite the unique status of the generally unquestioned supremacy enjoyed by the territoriality principle, it is unsurprising that we can find several examples of how the law seeks to mitigate the destructive effect that a strict application of the territoriality principle may have.

For example, cross-border data communications often transit servers in states located between the sender and the receiver. It is an undisputable fact that during such a journey, data is temporarily located in the territory of states other than the state of the sender and that of the receiver. Under a strict application of the territoriality principle, the states of the transit servers could claim jurisdiction.

Thus, for example, where a person in State A sends an e-book to a person in State C, she may be exposed to the jurisdiction of State B if the e-book's journey happens to pass through State B (a fact that the sender may, of course, neither necessarily, nor always, be able to predict in advance). If the content of the e-book is illegal in State B, the sender may, with no warning or real possibility to predict it, violate State B's law. Further, the owner of the server in State B may also be violating that State's laws. Such a situation is clearly undesirable, and several examples can be found of legal solutions put in place to avoid this destructive impact of a strict application of the territoriality principle. Here we may usefully point to Article 4(1)(c) of

[65] SS 'Lotus' (*France v Turkey*) 73.

Directive 95/46/EC: '[T]he controller is not established on Community territory and, for purposes of processing personal data makes use of equipment, automated or otherwise, situated on the territory of the said Member State, *unless such equipment is used only for purposes of transit through the territory of the Community*' (emphasis added). In the light of this, the step to accepting that data being temporarily, and perhaps randomly, stored on a server within a state (for example, due to a multistate cloud computing arrangement) should not always give that state jurisdiction over the data, may not be so dramatic after all.

To point to but one additional example, should one be needed, mention can be made of how the doctrine of *forum non conveniens* may work to mitigate the destructive impact of the territoriality-based ground for jurisdiction allowing the courts in Common Law countries to exercise jurisdiction based on the temporary presence of a person on the state's territory.[66]

In addition to this, it should be noted how our current territoriality focus is driving a movement that may have severe long-term implications for the Internet. I am here referring to the tendency of some states to make mandatory the localisation of data within their respective territories.[67] I need not, however, explore further here the topic of 'data localisation'—and the fragmentation, or Balkanisation, it brings.[68]

2.3.3 State Practice Has Already Moved Beyond Territoriality, Territorial Sovereignty, and the 'Harvard Draft' Principles

Khan has noted:

> [I]n recent years there are increasing signs that the traditional and rather categorical symbiosis between territory and power may no longer lay a legitimate claim for exclusivity. This is hardly deplorable since from an international law perspective, possession and transfer of territory have never been considered an end in itself. *L'obsession du territoire* of modern States was always meant to serve people, not vice versa.[69]

Several examples of a move in state practice away from territoriality, territorial sovereignty, and the 'Harvard Draft' principles can be found in the public

[66] See eg Dan Svantesson, *Private International Law and the Internet* (3rd edn, Kluwer Law International 2016) 180–202.

[67] See eg Russia's Federal Law No 242-FZ.

[68] See instead Daniel Cooper and Christopher Kuner, 'Data Protection Law and International Dispute Resolution' (2017) 382 Recueil des cours 9–174, 72–76.

[69] See eg Khan (n 46) 248 (footnote omitted).

international law arena. Consider, for example, the concept of humanitarian interventions about which Slaughter has noted:

Many international lawyers are deeply worried about the move away from the 'classicist' doctrine of non intervention. However, from a Liberal IR perspective, the legitimization of humanitarian intervention is a natural concomitant of human rights law. It is necessitated by some radical breakdown in the functioning of domestic institutions—the failure of a state to provide essential services such as food and shelter to its citizens or the active mass oppression of its citizens. In such circumstances, international action is justified to substitute for and in some cases even rebuild basic state institutions to the extent such efforts have a reasonable chance of success. Again, the justification is not moral, although it can certainly be, but instrumental, in terms of the likely impact of the humanitarian disaster on other states.[70]

In this context, we can also take note of what Slaughter describes as a shift from international to universal jurisdiction:

The International Criminal Court establishes direct individual criminal responsibility directly at the international level and uses a mixture of domestic and international tribunals to enforce it. Equally important is a more general effort to transform international jurisdiction into universal jurisdiction—to include more crimes in the category of *delicta juris gentium*, according to which states have an international obligation to try and punish certain conduct on the basis of their domestic criminal law.[71]

Ubertazzi has expressed similar thoughts:

[W]e are all witnessing a redeployment of the functions of the State and a redefinition of its relationship with its territory […] and therefore a tension exists between territorial sovereignty on the one hand, and the common interest of the international community on the other. This tension calls for the realisation of a fair balance between territorial and universal interests[.][72]

These practices also clearly mark a departure from strict territoriality, the 'Harvard Draft' principles as traditionally envisaged, and territorial sovereignty. Many other examples can be found. For example, Nagy provides an interesting discussion of how the 'borders' we rely on for immigration regulation are no longer aligned with state territories, concluding that 'the idea

70 Anne-Marie Slaughter, 'A Liberal Theory of International Law' (2000) 94 Am Soc Intl L Proc 240, 246.

71 ibid 247. For an interesting example of a proposal to expand the types of activity the International Criminal Court can deal with, see Steven Freeland, *Addressing the Intentional Destruction of the Environment During Warfare Under the Rome Statute of the International Criminal Court* (Intersentia 2015).

72 Benedetta Ubertazzi, *Exclusive Jurisdiction in Intellectual Property* (Mohr Siebeck 2012) 138.

that states' territories (and therefore: states themselves) are clearly separated, that where (the jurisdiction of) one state ends the other's starts, (or there lies an internationalised territory) is no longer valid'.[73]

In fact, even in the context of enforcement jurisdiction—the type of jurisdiction in relation to which territoriality has been held most sacred—we can see clear signs of the territoriality principle's stranglehold weakening. In the data privacy arena, for example, are found instances of mandatory standard contracts providing Data Protection Authorities (DPAs) the authorisation to conduct audits on the territory of other states. Kuner explains:

> The Spanish Data Protection Agency has conducted an audit of a third-party data processor located in Colombia regarding compliance with Spanish legal requirements for data transfers. It seems that the audit was part of an ongoing program to conduct audits of data processors outside the EU, and was carried out based on EU standard contractual clauses signed by the data importer which grant the DPA the same audit rights as it has in Spain. The Italian Data Protection Authority has also obtained the consent of Google to travel to California and audit the company's compliance with EU data protection law on its premises.[74]

At the minimum, this amounts to an undermining of—and indirect attack on—the territoriality principle in its most closely guarded setting: that of enforcement jurisdiction. And more generally, as noted by Cooper and Kuner: 'Data protection law routinely applies protections of local law to data processing operations in other countries.'[75]

Finally on this, in the context of public international law, we do well to recall that already in the 1927 *Lotus* decision, the view was expressed that 'there would seem to be a tendency towards a relaxation in the strict application of this principle [the territoriality principle].';[76] a sentiment echoed, for example, in the joint separate opinion of Judges Higgins, Kooijmans, and Buergenthal in *Democratic Republic of the Congo v Belgium (Arrest Warrant)*:

> As we said [...] a gradual movement towards bases of jurisdiction other than territoriality can be discerned. This slow but steady shifting to a more extensive application of extraterritorial jurisdiction by States reflects the emergence of values which enjoy an ever-increasing recognition in international society. One such value is the

[73] Boldizsár Nagy, 'From the National Border to the National Eleven: A (Partial and Partisan) Appraisal of the State System's Performance since the End of the Cold War' in James Crawford and Sarah Nouwen (eds), *Selected Proceedings of the European Society of International Law* (3rd vol, Hart 2012) 189.

[74] Kuner, 'Extraterritoriality and International Data Transfers in EU Data Protection Law' (2015) 5(4) Intl Data Privacy L 235–45, 240 (internal footnotes omitted).

[75] Cooper and Kuner (n 68) 9–174, 30.

[76] SS 'Lotus' (*France v Turkey*) (1927) PCIJ Series A, No 10, 63, per Nyholm.

importance of the punishment of the perpetrators of international crimes. In this respect it is necessary to point out once again that this development not only has led to the establishment of new international tribunals and treaty systems in which new competences are attributed to national courts but also to the recognition of other, non-territorially based grounds of national jurisdiction [...] what is regarded as a permissible jurisdiction [...] [is] in constant evolution.[77]

Interesting examples can also be found illustrating that, in part, the thinking of private international law now clearly goes beyond territoriality.[78] Indeed, Muir Watt goes as far as to claims that territory as defining jurisdiction was 'banished' in private international law long ago.[79] This is, however, an over-simplification and does not correspond well with reality. After all, as is widely known, several private international law rules are strongly territorial in nature, for example a contract formed within the territory may be a ground to exercise jurisdiction. Examples abound, and there is no need to list further examples here. Indeed, as pointed out by Crawford, the 'influence of the territoriality principle in private international law is [...] pervasive[.]'[80]

Nevertheless, Muir Watt's observations are valuable, and important examples can easily be found of private international law rules that seek to steer away from territoriality. Consider, for example, the 'centre of interests' or 'centre of gravity' test. While this test most famously, in recent times, was employed by the Court in the *eDate* case,[81] Oster emphasises that this test is not a recent invention and points to its use in the United States, Canada and by the House of Lords.[82]

[77] *Democratic Republic of the Congo v Belgium (Arrest Warrant)* [2002] ICJ Rep 3, paras 73–75.

[78] While a group of scholars no doubt would prefer to still treat jurisdiction under private international law as unrelated to jurisdiction under public international law, a soberminded consideration of the real state of things makes clear that such a position is untenable. I will, however, not enter that debate here.

[79] Horatia Muir Watt, 'A Private (International) Law Perspective: Comment on "A New Jurisprudential Framework for Jurisdiction"' (2015) 109 AJIL Unbound 75, 78. She also states: 'By the 1960s, the virtues of "mechanical jurisprudence" were unmasked during the U.S. "functionalist revolution" in the field. Since territory had been used to justify more or less any outcome—as the argument went—it could provide little guidance for solving conflicts of laws relating to complex torts; the proper reach of each conflicting claim was more plausibly defined, therefore, by policy concerns. In the early 1970s, the Second Restatement reflected this radical change from "power" to "policy": substantial connections, governmental interests, and more diffuse interests of interstate commerce or international legal order—the three guidelines advocated by Professor Svantesson—had replaced any reference to territoriality' (76).

[80] James Crawford, *Brownlie's Principles of Public International Law* (8th edn, OUP 2012) 474.

[81] Cases C-509/09 *eDate Advertising GmbH v X* and C-161/10 *Olivier Martinez and Robert Martinez v MGN Limited*.

[82] Jan Oster, 'Rethinking *Shevill*. Conceptualising the EU private international law of Internet torts against personality rights' (2012) 26 Intl Rev L, Compu & Tech 113, 120. To this, we may

Furthermore, he correctly links the test to Savigny's 'seat of the legal relationship'.[83]

Importantly, Oster also draws attention to the fact that the adoption of the 'centre of interests' test in the *eDate* case was made necessary by the failure of current—territoriality-focused—connections tests, such as the 'place where the information has been published'; this type of test, in Oster's assessment, 'is rendered almost valueless with a view to the global accessibility of online publications':[84]

> Instead of a determination based on objective facts such as physical publication of a newspaper as under *Shevill*, the 'centre of gravity of the conflict' doctrine introduces an evaluating interest analysis considering different criteria in order to establish the appropriate forum most suitable for the interests of all the parties, that is, essentially the most convenient or 'natural' forum.[85]

Perhaps we can go one step further and say that, rather than being 'valueless', the global accessibility of online publications makes tests, like the place where the information has been published, dangerous and vulnerable to abuse. One need only consider the wave of 'libel tourism'[86] cases to see that this is so. Where a court can claim jurisdiction based solely on the, often whimsical, fact that someone happened to read a defamatory article online while physically present in the territory where the court sits, there is an open invitation for libel tourism. This has sparked law reform, for example, in England where the result takes the form of the Defamation Act 2013.[87]

In the context of private international law, it is also of significance to note how the territoriality principle and the nationality principle have lost ground to other focal points such as 'domicile' and 'habitual residence'. In his interesting article 'Rethinking Jurisdiction in International Law', Mills highlights a highly significant consequence of this:

> The fact that these connecting factors are used and widely accepted in private international law itself suggests that the treatment of territory and nationality as discrete grounds for jurisdiction in traditional formulations of international law jurisdiction is too restrictive. The practice of states instead supports the idea that jurisdiction may be based on a flexible combination of both territorial and personal connecting

add its use in the various Australian Defamation Acts, such as in *Defamation Acts 2005* (Qld) s 11(2).

[83] ibid 120. [84] ibid 122. [85] ibid 121.

[86] See eg Matthew Collins, *Collins On Defamation* (OUP 2014) 502; James Price and Felicity McMahon (eds), *Blackstone's Guide to The Defamation Act 2013* (OUP 2013) 137.

[87] See eg Price and McMahon (n 86) 1.

factors – connections between a person and a place which do not depend on nationality, such as domicile or habitual residence.[88]

At least one leading commentator prefers to view the domicile principle as 'a variation on the active personality principle [another term for the nationality principle]'.[89] This may be so, but this does not necessarily undermine Mills' assertion that the focus on domicile and habitual residence highlights that state practice has moved beyond the nationality principle as originally conceived. And it may be argued that, if the domicile principle is to be viewed as a variation on the nationality principle, then the nationality principle has indeed been stretched beyond recognition.

It could, of course, also be argued that the move from nationality-focus to other focal points such as domicile and habitual residence in fact represent a strengthening of the territoriality principle, since both the concept of domicile and the concept of habitual residence are connected to place and thereby territoriality. However, both of those concepts have at least as much, I would say more, in common with a centre of interest analysis, as they do with the territoriality principle.

Finally, it is interesting to note that the focus on territoriality has been called into question even in relation to areas such as intellectual property rights that traditionally have been exclusively tied territoriality. For example, in her refreshing examination of jurisdiction in intellectual property, Ubertazzi concludes that:

> IPR [intellectual property rights] cannot be linked to a precise physical and geographical territory, but rather are social and universal phenomena. [. . .] [The] risk of transnational misappropriation of IPRs raises a number of issues of how to protect these rights universally, exposing the territorial principle to increasing doubts.[90]

All this highlights that state practice has, to a considerable extent, moved past a strict adherence to territoriality. Those inclined to start their sentences with the words 'Yes, but . . .' will no doubt counter that the territoriality principle nevertheless remains the cornerstone of jurisdiction and is adhered

[88] Mills (n 2) 207. See also Agata Jaroszek, 'European Online Marketplace—New Measures for Consumer Protection against "Old Conflict of Laws Rules"' (2015) 9 Masaryk UJL & Tech 23, noting that the European legislator has promoted the habitual residence concept 'as a connecting factor providing more legal certainty compared with other classical factors such as nationality and domicile.'

[89] Ryngaert (n 2) 108.

[90] Ubertazzi (n 72) 139 (footnotes omitted). For another interesting discussion of territoriality, and extraterritoriality, in the context of intellectual property law, see Alexander Peukert, 'Territoriality and Extra-Territoriality in Intellectual Property Law' in Günther Handl, Joachim Zekoll, and Peer Zumbansen (eds), *Beyond Territoriality* (Martinus Nijhoff 2012).

to more often that it is departed from. This I do not deny. My point is simply that we have successfully moved beyond strict territoriality in important respects and therefore that further areas of jurisdiction turning away from strict territoriality may be not only possible but indeed useful.

2.3.4 Territorial versus Extraterritorial

The terms 'extraterritorial' and 'extraterritoriality' are commonly used in the context of matters falling within the discipline of public international law. In contrast, one surprisingly rarely finds reference to those terms in writings on private international law (or 'conflict of laws'). Yet, at least on a practical level, extraterritoriality is as much an issue in private international law as it is in public international law. For example, where a court, in a civil matter, claims jurisdiction over a foreign defendant based on the fact that the defendant's writings have been read in the country claiming jurisdiction, it is of course making a jurisdictional claim with extraterritorial effect. Similarly, when a court in such a case decides to apply its domestic law to the foreign conduct by the foreign party, it is giving extraterritorial effect to its law.

In fact, the centrality of the role played by extraterritoriality is such that it is difficult to think of any private international law case of note that does not involve some dimension of extraterritoriality. For example, it is often said that the Internet gives rise to controversial jurisdictional issues. However, there is nothing remarkable or controversial about, for example, a country exercising jurisdiction over an Internet dispute involving two of its citizens who have acted within that country. Indeed, such a situation can typically be dealt with in the same way whether the parties used the Internet or not. The truth is that it is only where a country seeks to exercise jurisdiction over Internet conduct *in an extraterritorial manner* that the remarkable features of the Internet become relevant and that controversies arise. In other words, extraterritoriality is the key ingredient in every controversial claim of jurisdiction in relation to the Internet.

While it may be tempting to suppose that extraterritoriality is a modern phenomenon, this is not so. In fact, extraterritoriality has a long history indeed:

> In seeking the origin of extraterritoriality, some jurists and historians trace it to the imperialistic period of the last century. Others find the origin of this jurisdiction in the 'letters of privilege' which the Greek Christian rulers at Constantinople, and later their Moslem conquerors, issued to the city republics of Italy in the 11th and 12th centuries. Again, others trace it to the period of the Roman Empire. Quite a few writers, on the other hand, seem to be satisfied with the obviously less troublesome

> assumption, that the original document concerning extraterritoriality is to be found in the treaty of 1535 between the Franks and the Turks.
>
> It is the writer's opinion that the origin of extraterritorial privileges has to be traced much earlier even than the period of early Rome and the Germanic tribes. Traces are found in the more ancient world. The principle of territorial law and sovereignty was unknown in the ancient world. It was even vague in the Middle Ages. In fact, kingdoms during the medieval period had vague and uncertain boundaries. Indeed, even during the first centuries of what we call modern history, no such conceptions as territorial law or territorial sovereignty were entertained. Sovereignty was not associated with dominion over a territory. It was a tribe-sovereignty. [. . .] Early man stood in great fear of the magic of strangers. He resorted to a variety of ceremonies in order to protect himself against the devilish power of the stranger. But in the course of time the natives were convinced that this attitude of strict exclusiveness could not be permanently maintained. As soon as they failed to find in their own association the satisfaction of their desires and the supply of their wants, they were compelled to go beyond it and enter into relations of some sort with the surrounding world. The alien, therefore, was incapable of amenability to the same jurisdiction to which the natives were subjected. For this reason we find that in the ancient world foreigners were either subjected to their own laws and customs or were placed under a special jurisdiction. It is in these relations and under these conditions that we find the earliest traces of extraterritoriality.[91]

In any case, put simply, convention would have us believe that jurisdictional claims are either territorial or extraterritorial, with the latter type often described as relating to 'the exercise of jurisdiction by a State over activities occurring outside its borders'.[92] And one of the key consequences of the centrality of the territoriality principle is that a great deal of focus has been directed at distinguishing between situations involving 'territorial' jurisdictional claims from jurisdictional claims that are 'extraterritorial'.[93] Yet it seems clear that to mark the difference in this way is to invite confusion, and there are two serious problems with this distinction.

As noted by Ireland-Piper: '[D]omestic laws with extraterritorial scope do not fit neatly in either domestic or international law frameworks and raise issues under both. In this way, there is a third space, the regulation of which

[91] Shalom Kassan, 'Extraterritorial Jurisdiction in the Ancient World' (1935) 29 AJIL 237, 239–40 (footnotes omitted).

[92] Senz and Charlesworth (n 57) 72.

[93] The International Law Commission has defined 'extraterritorial jurisdiction' as 'an attempt to regulate by means of national legislation, adjudication or enforcement the conduct of persons, property or acts beyond its borders which affect the interests of the State in the absence of such regulation under international law' (International Law Commission, *Report on the Work of its Fifty-Eighth Session* (1 May–9 June and 3 July–11 August 2006) UN Doc A/61/10, Annex E, para 2).

lacks clarity.'[94] In other words, there are obvious difficulties in categorising the extraterritorial scope of domestic law. Unlike Ireland-Piper, I am more inclined to view this as an issue for both domestic and international law rather than as something falling outside both domestic and international law. More importantly, however, I fully agree that there is a lack of clarity as to the regulation of domestic laws with extraterritorial scope.

Indeed, too often it is not possible to draw a distinction between 'territorial' and 'extraterritorial' jurisdictional claims. A relatively recent article by Scott provides a great illustration of this problem. She discusses a set of new jurisdictional 'triggers' in European Union law and seeks to view them through the lens of the traditional distinction between what is territorial and what is extraterritorial. Because of the flaw in that distinction, she frequently has to conclude that 'the distinction between territorial and extraterritorial legislation is often unclear'.[95] Examples proving that this is so abound. Consider, for example, the question of whether a state is exercising jurisdiction 'over activities occurring outside its territory' where it regulates the use of personal information about its citizens stored in a cloud computing arrangement with multi-jurisdictional reach. Or consider the already mentioned Article 4(1)(c) of Directive 95/46/EC of the European Parliament and of the Council of 24 October 1995 on the protection of individuals with regard to the processing of personal data and on the free movement of such data.[96]

The fact that this provision is based on the territoriality principle need not be debated; it is obvious from the fact that it is focused on the territory on which a controller makes use of equipment. However, as the saying goes, 'the devil is in the detail', and just how easily we may be misled is obvious when we consider the fact that just about anything can be regarded as 'equipment'. For example, the highly influential Article 29 Working Group has stated that 'cookies' placed on a computer in the European Union may amount to 'equipment'.[97] Surely that would justify viewing this Article as extraterritorial in reach, under the territorial versus extraterritorial distinction? Or would the fact that the language used in the Article is focused on the use

[94] Danielle Ireland-Piper, *Accountability in Extraterritoriality* (Edward Elgar 2017) 5.

[95] Joanne Scott, 'The New EU "Extraterritoriality"' (2014) 51 Common Mkt L Rev 1343. One need only read the abstract to find statements such as: 'EU legislation that relies on such triggers is neither self-evidently territorial nor extraterritorial, and it remains unclear whether it is consistent with the territorial principle of jurisdiction' (1343). Along the same lines, see also eg 'Where the EU has recourse to novel triggers of this kind, the line between territorial jurisdiction and extraterritorial jurisdiction becomes difficult to draw' (1345).

[96] OJ (L 281).

[97] Article 29 Working Party, *Opinion 2/2010 on online behavioural advertising* (22 June 2010) <http://ec.europa.eu/justice/policies/privacy/docs/wpdocs/2010/wp171_en.pdf>, 21, accessed 2 April 2017.

of equipment on the relevant territory mean that this Article is not extra-territorial in nature? As Crawford cautiously—perhaps overcautiously—puts it: '[W]hat amounts to extra-territorial jurisdiction is to some extent a matter of appreciation.'[98]

This tells us something important; there is an undeniable tension between, on the one hand, the legal framework we apply with its attempt at a sharp distinction between what is territorial and what is extraterritorial and, on the other hand, reality. And where such a tension exists it would seem prudent to prima facie assume that it is the legal framework that needs adjustment.

Furthermore, even if we were able to draw a sharp line between jurisdictional claims that are territorial and those that are extraterritorial, identifying a jurisdictional claim as being extraterritorial tells us little, or nothing, of value. Some extraterritorial claims can be indisputably legitimate and useful (after all they may be based eg on the widely recognised nationality principle), while other extraterritorial claims are equally indisputably illegitimate and excessive.[99] Yet too often the ill-advised territorial/extraterritorial distinction is used as shorthand for legitimate (ie territorial) claims versus illegitimate (ie extraterritorial) claims of jurisdiction.[100] Such oversimplifications are misguided and unhelpful, and invariably create obstacles for a fruitful debate. Indeed, in his insightful book *Jurisdiction in International Law*, Ryngaert concludes that, because the term 'extraterritorial' is tainted by pejorative connotations acquired over the years, the term ought to be avoided.[101] I wholeheartedly agree.

[98] Crawford (n 80) 457.

[99] See also Dan Svantesson, 'Extraterritoriality and Targeting in EU Data Privacy Law: The Weak Spot Undermining the Regulation' (2015) 5(4) Intl Data Priv L 226–34; Kuner (n 74) 235–45.

[100] For an example of such a point of view, see Joanne Scott, 'The New EU "Extraterritoriality"' (2014) 51 Common Mkt L Rev 1343, 1345: 'While it is plain that the EU legislation in question regulates activities that take place abroad, what is not straight forward is the determination of whether the triggers that the EU is relying on should be understood as territorial or not. This is important because although the concept of extraterritoriality is often used as a slogan to cast doubt on the legitimacy of a measure without unpacking what exactly this means, the concept of territory has clear legal significance as well. Indeed, the territoriality principle is one of the most important bases for the exercise of prescriptive jurisdiction recognized by customary international law. Consequently, where a measure is defined as extraterritorial, it will be unlawful unless an alternative, recognized jurisdictional base can be found' (footnote omitted).

[101] Ryngaert (n 2) 8. See also Kuner stating that: '[b]eyond the disagreements about its meaning as a legal concept, the term has acquired political baggage that makes its use charged with emotion. Some view the extraterritorial application of data protection law positively as a mechanism for protecting the rights of individuals from threats outside the jurisdiction, while others see it as an improper intrusion of foreign states on domestic interests' (Kuner (n 74) 235–36 [footnotes omitted]).

So, to bring this section to a conclusion, the law has tried and failed to draw, and maintain, a meaningful distinction between what is territorial and what is extraterritorial. This venture was worthwhile, but it has failed and we should now accept this failure and move on; a continued use of this distinction is both a source, and a sign, of confusion.

The question is then, what we should move on to? Back in 2013,[102] I expressed the view that we may turn our focus to whether or not a jurisdictional claim has 'extraterritorial effect' or not; that is, an assertion of jurisdiction ought to be regarded as extraterritorial as soon as it seeks to control or otherwise directly affect the activities of an object (person, business, etc.) outside the territory of the state making the assertion.[103]

This proposal overcomes the first of the two problems outlined above—it makes it feasible to draw a reasonably sharp line between what is territorial and what is extraterritorial. However, I now doubt that this definition can fully overcome the second problem I sought to bring attention to above.[104] After all, in many, if not a majority, of cases a claim of jurisdiction will have some extraterritorial effect as defined above. Thus, focusing on an extraterritorial effect will presumably not satisfy the threshold for extraterritoriality, for example, in the context of the presumption against extraterritoriality found in the laws of many countries. As pointed out by the US Government in the *Microsoft Warrant* case[105] discussed on multiple occasions in the text: 'The principle against extraterritoriality presumes that Congress does not intend for a law to apply extraterritorially. It does not presume Congress's intention to be that the law has no incidental effects outside the country whatsoever.'[106]

In the light of the this, our obsession with the territorial/extraterritorial divide must come to an end. However, I recognise that such a change will not happen overnight. Just how deeply rooted our obsession with territoriality is becomes clear when we consider how legal instruments, such as Australia's

[102] See eg Dan Svantesson, *Extraterritoriality in Data Privacy Law* (Ex Tuto 2013) 85.

[103] See also Yves Poullet, 'Transborder Data Flows and Extraterritoriality: the European Position' (2007) 2 J Intl Compu L & Tech 141, 148, distinguishing between 'extraterritorial impact' on the one hand, and 'extraterritorial scope' on the other.

[104] See also Kuner (n 74) 235–45.

[105] *In the Matter of a Warrant to Search a Certain E-mail Account Controlled and Maintained by Microsoft Corporation*, No 14-2985, 2016 WL 3770056 (2d Cir 14 July 2016).

[106] Government's Brief in Support of the Magistrate Judge's Decision to Uphold a Warrant Ordering Microsoft to Disclose Records Within its Custody and Control, *In the Matter of a Warrant to Search a Certain E-mail Account Controlled and Maintained by Microsoft Corporation* (1:13-mj-02814) at 19. See also *Environmental Defense Fund v Massey* 986 F2d 528, 531–32 (DC Cir 1993): 'Even where the significant effects of the regulated conduct are felt outside U.S. borders, the statute itself does not present a problem of extraterritoriality, so long as the conduct which Congress seeks to regulate occurs largely within the United States.'

Crimes (Child Sex Tourism) Amendment Act 1994 (Cth), invariably are discussed as being 'extraterritorial'.

The fact that laws such as these are extraterritorial is of course as uninteresting as it is uncontroversial. All that observation tells us is that the law operates beyond the territorial limits and therefore cannot be legitimised by reference to the territoriality principle; it is thus merely a negative statement excluding reliance on the territoriality principle. The considerably more interesting question of what principle is in fact used instead to legitimise such a jurisdictional claim is obscured by this obsession with territoriality.

As is well known, the jurisdictional claims made through instruments such as the *Crimes (Child Sex Tourism) Amendment Act 1994* (Cth) are, apart from being legitimised by treaty obligations, legitimised by reference to the nationality principle—that is, what is important is not the grounds *not* relied upon, such as territoriality. Indeed, it is only where we do not rely on the territoriality principle that we engage in the odd conduct of focusing on jurisdictional grounds not relied upon. After all, when a state claims jurisdiction over a non-national based on activities that person undertook within that state's territory, we do not speak of that as, for example, an extra-national jurisdictional claim.

One last aspect of the territorial/extraterritorial distinction ought to be noted. While we never seem to question how we define territoriality, if territorial jurisdiction is the opposite of extraterritorial jurisdiction, and—as has been shown—we cannot define extraterritorial jurisdiction, then how can we imagine that we have a clear definition of territorial jurisdiction?[107] This definitional conundrum also undermines the territoriality principle.

The main conclusion to be drawn from the above is this: if, as has been demonstrated, the conceptual distinction between territoriality and extraterritoriality does not fully work, the usefulness of the territoriality principle is

[107] The risks associated with adopting the conventional platitudes in this field are well illustrated in Colonna's otherwise excellent 'Legal Implications of Data Mining: Assessing the European Union's Data Protection Principles in Light of the United States Government's National Intelligence Data Mining Practices' (Ragulka 2016). In discussing the difference between territorial jurisdiction on the one hand, and extraterritorial jurisdiction on the other, Colonna states: 'Defining the concept of territorial jurisdiction is relatively straightforward. [...] Unlike territorial jurisdiction, the concept of extraterritorial jurisdiction is much more elusive and defining it within the context of data protection is fraught with difficulties' (ibid 325). At the same time she claims that: 'all jurisdiction that is not exclusively territorial can be thought of as extra territorial jurisdiction' (ibid 326). However, if defining the concept of territorial jurisdiction is indeed relatively straightforward and all jurisdiction that is not exclusively territorial can be thought of as extraterritorial jurisdiction, then it is of course a logical impossibility that the concept of extraterritorial jurisdiction is difficult to define.

severely undermined and it is no longer deserving of the supreme position it holds currently.

2.3.5 Territoriality as a Component of a Dysfunctional System

Early debates about Internet jurisdiction centred on the argument that it is difficult, or even impossible, to identify the location of online activities. While we need to continue that discussion for example in the context of cloud computing, as the debate has matured the greater concern is now that many online activities, intentionally or unintentionally, touch upon the territories of states without having any real substantial connection to those states. I think the discussions of these problems can usefully act as a catalyst for broad reform of how international law approaches jurisdiction; after all as noted by Muir Watt:

> Territory did not wait for the internet, either, to prove inadequate in defining jurisdiction in civil cases with criminal overtones, where geography clearly counted less and less: money laundering, financial fraud, trafficking (in cultural objects, children or human organs) joined more familiar transnational 'toxic torts' and cross-border invasions of privacy, at the nexus between the public and the private, soliciting the resources of both branches of international law. The coming of cybertorts, with cultural values colliding dramatically in global virtual space, only confirmed the need for a methodological overhaul.[108]

Thus, the reform called for is required in relation to legislative jurisdiction, adjudicative jurisdiction, and enforcement jurisdiction as well as what I elsewhere describe as a fourth category— that of investigative jurisdiction— both in and beyond the online context.

However, before moving on to consider possibilities for a better approach to jurisdiction, it is worth noting that the territoriality principle and the other 'Harvard Draft' principles operate within a system—a dysfunctional system—with additional components such as (territorial) sovereignty, comity, due diligence, and the duty of non-interference.

Many readers will no doubt be familiar with these concepts. Yet, partly for convenience and partly due to the existing variations in how these concepts are defined, I will devote a few lines to discussing their respective definitions here.

As to the term 'sovereignty',[109] it has been said, 'Among the theories of jurists, there is, perhaps, none which has been a battle-ground for so

[108] Muir Watt (n 79) 76.

[109] For an interesting discussion of 'sovereignty' as a cliché, see Karen Knop and Susan Marks, 'The War against Cliché: Dispatches from the International Legal Front' in Christine Chinkin

long a time, as that which relates to the limits of sovereign power.'[110] To avoid unnecessary complexity, I think it appropriate here to focus predominantly on territorial sovereignty, and we can safely embrace this definition of the concept: 'Sovereignty in the relations between States signifies independence. Independence in regard to a portion of the globe is the right to exercise therein, to the exclusion of any other State, the functions of a State.'[111] However, in a broader sense, Lauterpacht's observation is informative:

> [T]he impact of the theory of the sovereignty of States reveals itself in the international sphere mainly in two ways: first, as the right of the State to determine what shall be for the future the content of international law by which it will be bound; secondly, as the right to determine what is the content of existing international law in a given case.[112]

While this broader aspect of sovereignty has a less direct impact on jurisdiction, it is nevertheless of fundamental importance. After all, it impacts *how* jurisdictional rules may evolve.

More importantly for the arguments advanced in this book, we must note that orthodox thinking seems to conflate sovereignty and jurisdiction beyond what is reasonable:

> The position that states can assert or exercise jurisdiction only in relation to matters and persons within their sovereign realm is unsustainable as a general theory of jurisdiction under public international law. It is not obvious at all that the boundary between national jurisdictions runs in the same place as the boundary between national sovereignties. It can be said with certainty that there can be no more than one sovereign over a particular territory. It could not be said with the similar certainty that any single person or matter should always and exclusively be subject to one single sovereign power. In other words, sovereignty does not do the same job in relation to the boundary between jurisdictions as it does in relation to the boundary between state territories. Territorial supremacy of the state does not determine any template for the ultimate reach of, or limitations on, that state's jurisdiction.[113]

and Freya Baetens (eds), *Sovereignty, Statehood and State Responsibility: Essays in Honour of James Crawford* (CUP 2015) 17–18.

[110] A Lawrence Lowell, 'The Limits of Sovereignty' (1888–89) 2 Harv L Rev 70, 70.

[111] *Island of Palmas (Netherlands v US)* 2 RIAA 829, 838 (Perm Ct Arb 1928). This definition is also adopted in Schmitt (n 43) 16.

[112] Hersch Lauterpacht, *The Function of Law in the International Community* (OUP 2011) 3.

[113] Alexander Orakhelashvili, 'State Jurisdiction in International Law: Complexities of a Basic Concept' in Alexander Orakhelashvili (ed), *Research Handbook on Jurisdiction and Immunities in International Law* (Edward Elgar 2015) 10.

The concept of comity does not lend itself to being easily pinned down. As a result, there are both divergent definitions and divergent views of the value of comity. To avoid departing too far from the main theme, here it will have to suffice to note that arguably the most widely used definition would have us view comity in these terms:

Comity in the legal sense, is neither a matter of absolute obligation on the one hand nor of mere courtesy and good will upon the other. But it is the recognition which one nation allows within its territory to the legislative, executive or judicial acts of another nation, and to the rights of its own citizens or of other persons who are under the protection of its laws.[114]

The duty of non-intervention, or the principle of non-interference, is in a general sense nothing but the other side of the sovereignty coin. As noted by Gerber:

When a state attaches legal consequences to conduct in another state, it exercises control over that conduct and when such control affects essential interests in the foreign state, it may constitute an interference with the sovereign rights of that foreign state. Consequently, the principle of non-interference is properly applicable to the exercise of jurisdiction.[115]

Nevertheless, the principle of non-interference has come under fire and its current and future role, if any, in international law appears unsettled and debated.[116] Furthermore, if applied strictly, it is also clearly incompatible with some common jurisdictional claims:

[114] *Hilton v Guyot* 159 US 113 (1895) 164. For a more elaborate discussion of the concept of comity, see eg Adrian Briggs, 'The Principle of Comity in Private International Law' (2012) 354 Recueil des Cours 65, 94, as well as two highly interesting articles by Thomas Schultz: Thomas Schultz and Jason Mitchenson, 'Navigating Sovereignty and Transnational Commercial Law: The Use of Comity by Australian Courts' (2016) 12(2) J Priv Intl L 344–78 <http://dx.doi.org/10.1080/17441048.2016.1206704>, accessed 2 April 2017; Thomas Karl Peter Schultz and Niccolò Ridi, 'Comity: The Narrative and Influence of a Core American Concept' (2017) 9(2) Northeastern U LJ.

[115] Gerber, 'Beyond Balancing', 212. This thinking is supported by Crawford:

'The corollaries of the sovereignty and equality of states are: (a) a jurisdiction, prima facie exclusive, over a territory and the permanent population living there; (b) a duty of non-intervention in the area of exclusive jurisdiction of other states; and (3) the ultimate dependence upon consent of obligations arising whether from customary law or from treaties' (Crawford (n 79) 457). The Declaration on Principles of International Law Concerning Friendly Relations and Co-operation among States in Accordance with the Charter of the United Nations, A/RES/25/2625, 1970 affords a more technical meaning to the 'duty not to intervene in matters within the domestic jurisdiction of any State', but a discussion of the details of this goes beyond the present scope.

[116] See eg Chatham House, 'The Principle of Non-Intervention in Contemporary International Law: Non-Interference in a State's Internal Affairs Used to Be a Rule of International Law: Is It Still?' (28 February 2007).

Pursuant to the principle of non-intervention, States are prohibited from intervening in the domestic affairs of other States. If this principle were to be strictly applied in the law of prescriptive jurisdiction, it may give rise to a finding of any jurisdictional assertion that reaches beyond a State's boundaries being in violation of international law.[117]

The due diligence principle, and the related and overlapping 'no harm principle', is perhaps primarily associated with the well-known 1949 *Corfu Channel* case where the International Court of Justice (ICJ) stressed 'every State's obligation not to allow knowingly its territory to be used for acts contrary to the rights of other States.'[118] However, it has a longer history, and in his dissenting opinion in the *Lotus* case, Moore refers to *United States v Arjona* (1887) 120 US 479 and observes: 'It is well settled that a State is bound to use due diligence to prevent the commission within its dominions of criminal acts against another nation or its people[.]'[119] And more importantly, the dissenting opinion of Weiss pointing to the correlation between sovereignty, jurisdiction and due diligence stated:

> By virtue of sovereignty such as we understand it, every State has jurisdiction to sentence and punish the perpetrators of offences committed within its territory; indeed, this is a question of public security, and of public order, which a State cannot ignore without neglecting its duty as a State, and one which arises whatever the nationality of the delinquent may be.[120]

At any rate, while this principle is primarily discussed in the context of international environmental law, international human rights law, international humanitarian law, and international investment law, its prominence in the Cyber world is clearly increasing, as is indicated by the plentiful recent literature on the topic.[121] Indeed, several states—such as Germany—have expressly started to implement the due diligence principle in their Cyber policies.[122]

Given these definitions, the concepts and principles discussed form a system with an endless, and I argue insufferable, number of clashes, contradictions, and logical incoherence. Amongst these clashes and contradictions, one stands out above the rest: where we embrace either objective territoriality,

[117] Ryngaert, *Jurisdiction in International Law*, 154. [118] [1949] ICJ Rep 4, 22.

[119] SS 'Lotus' (*France v Turkey*), 89. [120] ibid 44.

[121] See eg Michael N Schmitt, 'In Defense of Due Diligence in Cyberspace' (22 June 2015) 125 Yale LJ Forum 68 <https://ssrn.com/abstract=2622077>, accessed 8 April 2017.

[122] Annegret Bendiek, 'Due Diligence in Cyberspace: Guidelines for International and European Cyber Policy and Cybersecurity Policy' (Stiftung Wissenschaft und Politik -SWP- Deutsches Institut für Internationale Politik und Sicherheit, ed, 2016) SWP Research Paper 7/ 2016 <http://nbn-resolving.de/urn:nbn:de:0168-ssoar-47152-8>, 31, accessed 2 April 2017.

passive personality, or the effects doctrine, we quite simply have to reject territorial sovereignty, at least in its strictest forms.

Indeed, it seems to me that there is not one single proposed solution to the issue of Internet jurisdiction that cannot both be supported, and at the same time objected to, by reference to public international law principles such as territorial sovereignty, comity, duty of non-intervention, and the jurisdictional principles advanced in the 'Harvard Draft'. Such is the state of international law as it relates to jurisdiction.

Consider, for example, the currently ongoing data privacy versus law enforcement dispute between Microsoft and the US Government.[123] In December 2013, the US Government served a search warrant on Microsoft under the *Electronic Communications Privacy Act* of 1986. The warrant authorises the search and seizure of information associated with a specified web-based email account that is stored at Microsoft's premises. Microsoft has opposed the warrant since the relevant emails are located exclusively on servers in Dublin, Ireland. A key question in the matter is whether the United States would be engaging in extraterritorial law enforcement in Ireland where the data sits, even though all actions taken to retrieve that data would be taken from the United States.

Here, Microsoft sits as the proverbial 'meat in the sandwich', and the real dispute is between the United States, on the one hand, and Ireland/the European Union on the other hand. And, as Ryngaert discusses in detail, such trans-Atlantic disputes have some history in the setting of orders for discovery abroad.[124]

The United States may clearly argue that the territoriality principle supports its claim. It may perhaps even argue that Irish objections to the release of the relevant information violates Ireland's duty of non-intervention[125] and interferes with US sovereignty. At the same time, and with at least equal force, Ireland may argue that the US warrant violates its sovereignty and may also point to the territoriality principle, the doctrine of comity, and possibly the duty of non-interference. Other examples of situations in which the international law system as described does not provide answers abound. For example, in the work on the first *Tallinn Manual on the International Law*

[123] I have discussed this case in more detail elsewhere: Dan Svantesson and Felicity Gerry, 'Access to Extraterritorial Evidence: The Microsoft Cloud Case and Beyond' (2015) 31 Compu L & Sec Rev 478–89. On 23 June 2017 the US Department of Justice petitioned for the matter to be heard by the US Supreme Court.

[124] Ryngaert, *Jurisdiction in International Law*, 89–94.

[125] Mills describes the territorial character of enforcement jurisdiction as 'an important reflection of the principle of non-intervention in the internal affairs of other states': 'Rethinking Jurisdiction in International Law', 195.

Applicable to Cyber Warfare, it was noted that '[t]he International Group of Experts could achieve no consensus as to whether the placement of malware that causes no physical damage [...] constitutes a violation of sovereignty.'[126] It is fair to say that where a group of our most distinguished experts cannot agree on a matter such as this, that disagreement signals a failure of the relevant norms.

International law simply does not tell us where the conduct in question takes place in situations such as that of the *Microsoft Warrant* case,[127] and it could only ever do so by arbitrarily adopting a controversial legal fiction. In other words, given that we cannot always logically define where events online take place, the territoriality principle would not help even if it was free from defects in a general sense. This situation is nonsensical, and the time has surely come to start over.[128]

The mentioned dispute between Microsoft and the US Government may also be used to illustrate the definitional difficulties associated with drawing a line between what is territorial and what is extraterritorial, discussed earlier. Microsoft says the issue of extraterritoriality obviously does arise, and the US Government claims that it equally obviously does not. The difference in perspective is apparent throughout but is particularly well illustrated in this quote from the Government's brief of 9 June 2014:

> Relying on Section 432(2) of the Restatement (Third) of Foreign Relations, Microsoft argues that '[a] state's law enforcement officers may exercise their functions in the territory of another state only with the consent of the other state.' [...] But requiring the disclosure of records by a U.S. company does not involve any enforcement activity by government personnel on foreign territory, which is the concern of that section.[129]

[126] Schmitt (n 43) 16.

[127] *In the Matter of a Warrant ... Microsoft Corporation* (n 106).

[128] See also Hannah L Buxbaum, 'Territory, Territoriality, and the Resolution of Jurisdictional Conflict' (2009) 57 Am J Comp L 631, 635 noting:

> '"Territoriality" and "extraterritoriality," though, are legal constructs. They are claims of authority, or of resistance to authority, that are made by particular actors with particular substantive interests to promote. Thus, those harmed by anti-competitive conduct may invoke the concept of 'objective territoriality' to argue that the location of that conduct's effects within a particular country is sufficient to give that country jurisdiction to regulate; the government of a nation in which defendant corporations are located, conversely, may invoke the concept of "extraterritoriality" to argue that the regulating country would overstep its authority if it applied its laws to foreign conduct.'

[129] Government's Brief in Support of the Magistrate Judge's Decision to Uphold a Warrant Ordering Microsoft to Disclose Records Within its Custody and Control, *In the Matter of a Warrant to Search a Certain E-mail Account Controlled and Maintained by Microsoft Corporation* (1:13-mj-02814) 21.

Reading this quote carefully, it is obvious that Microsoft and the US Government are talking about two different things, and that they are arguably both correct. It is true, as the Government says, that there is no enforcement activity on foreign territory. However, and this is important, there is an exercise of law enforcement functions in the territory of another state. In other words, the Government looks exclusively to the location from which jurisdiction is exercised (the United States). Microsoft also considers the extraterritorial effects and those effects occur in Ireland. In this way, the US Government gives extraterritoriality a narrow definition, while Microsoft gives it a broader definition.

2.3.6 Not Just an Online Issue

The discussion has been centred on the proposition that the territoriality thinking that is so characteristic of our current paradigm is a bad fit for Cyberspace. But it is also increasingly obvious that the territoriality principle is a bad fit for the 'real world'. To see that this is so, we need only consider areas such as environmental law, air and space law,[130] anti-trust law, and the effective enforcement of human rights law. In all those fields, a strict adherence to the territoriality principle is problematic, and indeed directly harmful. So, the good news, if it can be seen to be news, is that we do not need to show that the online environment is different; the online environment is just one more illustration of the problems with the territoriality principle as such.

The implications of this are significant. First, it means that we may boldly aim to reform our rules of jurisdiction generally, not just as they apply online. Second, if it is felt that *sui generis* rules for jurisdiction online are the better option, it can be anticipated that the reform made in that regard may usefully assist reform initiatives in other fields of law.

The discussion of the offline problems with territoriality can also be approached from a different angle. One potential reason why people are so comfortable with territoriality thinking is that parallels to state territoriality can be found in our everyday lives; in other words, the territoriality focus is 'conceptually scalable' within the various levels of law (local, domestic, regional, and international). We need only consider our relationships with our neighbours to see that this is so. In the majority of cultures, we are today

[130] Consider eg the complications that have arisen in the context of the partly territoriality-based definitions of the 'launching state' (Art VII of the Treaty on Principles Governing the Activities of States in the Exploration and Use of Outer Space, including the Moon and Other Celestial Bodies 1967 and Art I of Convention on International Liability for Damage Caused by Space Objects of 1972).

used to the idea that our yard is separated from the neighbour's yard by a dividing line—a territorial border. We are also used to the idea that the location of activities and things may be acceptable or not acceptable depending where they take place in relation to that territorial border. I can pick the mangoes that grow on the trees in my yard because the mango tree stands in my yard, and I can kick a football around on my yard because I (and hopefully the ball) remain on my side of the territorial border.

But what happens if a branch of my mango tree grows across the territorial border into the neighbour's yard, if my football skills are such that the ball accidentally ends up in the neighbour's yard or, indeed, if the neighbour starts playing loud music at 2 o'clock in the morning? Clinging on to a strict territoriality thinking in such a scenario would presumably mean that (1) the neighbour has the right to spray poison on the branch of my mango tree that happens to have entered his yard, (2) that he has the right to keep my football, and (3) that there is nothing I can do about the loud music interrupting my sleep. Fortunately, that is not how our world is organised.

In the type of neighbour disputes outlined, I can, of course, rely on laws with common applicability across the territorial border between my neighbour and me, a court with jurisdiction over us both, and an enforcement body (the police) with power over us both. This is how we solve the disputes that arise due to our territoriality thinking, on the most local level. Unfortunately, however, this solution is not at all scalable to the international context.

2.4 Concluding Remarks Regarding Territoriality

It is interesting to consider how globalisation has resulted in an increased focus on territoriality in some areas. Consider, for example, the current strong focus on the use of local produce in the food arena, perhaps best illustrated by the movement described as *New Nordic Cuisine*.[131] Almost all of the ten points in the Nordic Cuisine 'Manifesto'[132] relate, one way or another, to territoriality.[133]

With law it is, or at least ought to be, quite the opposite. Globalisation impacts law by forcing us to become more aware of, and indeed to learn from,

[131] Nordic Council of Ministers, *New Nordic Cuisine* (2008) http://nynordiskmad.org/fileadmin/webmasterfiles/PDF/Ny_Nordisk_Mad_Low.pdf, accessed June 2017.

[132] ibid.

[133] Point 3 is an illustrative example: 'To base cooking on raw materials which characteristics are especially excellent in our climate, landscape and waters' (ibid 15).

foreign law and legal traditions. The increased international contacts that globalisation brings, not least through the Internet, also put stress on traditional legal notions such as territorial sovereignty—where so many activities have cross-border dimensions it makes little sense to place so much significance on where events occur. Thus, while globalisation leads to an emphasis on territoriality in areas such as food culture, in law, globalisation distinctly undermines territoriality.

Having provided several historical examples, Khan asserts: 'It seems in fact that the ingredients of a conflict over territory or boundaries, the desire to overcome it by the most accurate drawing of lines, and, finally, the means to achieve this end have not changed much in the course of the law four and a half millennia.'[134] This is both highly interesting and incorrect (or at least incomplete). It is incorrect (or at least incomplete) as it is clear that Cyber conflicts, as well as non-Cyber conflicts with Cyber elements, prove that we are now witnessing a dramatic change. It is interesting in that Khan's observation highlights just how significant is this change—it is a change that runs counter to the tradition of four and a half millennia.

In this context, it is instructive to note how '[t]echniques of mapping facilitate the abstraction of territory from the physical earth and allow the concept of territory to act as a mediator between sovereignty/jurisdiction and the physical,'[135] and modern cartography has been described as the 'technological midwife' of territorial sovereignty.[136] In other words, there is a clear correlation between the prominent role of territoriality and the development of mapping of our physical environment. As illustrated by the works of distinguished geographers,[137] also the online environment can be 'mapped' in many interesting ways. However, from a legal perspective the online environment maps in a different way to the physical world. Thus, the territoriality principle—the central pillar of the 'Harvard Draft' principles, and an argued

[134] Khan (n 46) 230.

[135] Shaunnagh Dorsett, 'Mapping Territories' in Shaun McVeigh (ed), *Jurisprudence of Jurisdiction* (Routledge-Cavendish 2007) 149.

[136] Ford (n 4) 873. Ford notes that '[t]erritorial identities developed and matured along with the advance of modem, scientific cartography. Once cartography made the production of precisely demarcated legal territories possible, territorial relationships quickly became dominant' (845) and that '[a]lthough cartography in ancient civilizations may have been fairly sophisticated, modem scientific cartography was not practiced in Europe until, at earliest, the fifteenth century' (873).

[137] See works such as Georg Glasze, 'Geoinformation, Cartographic (Re)Presentation and the Nation State: A Co-Constitutive Relation and Its Transformation in the Digital Age' in Uta Kohl (ed), *The Net and the Nation State—Multidisciplinary Perspectives on Internet Governance* (CUP 2017) 202–17, and Barney Warf, 'Alternative Geographies of Cyberspace' in Kohl (ed), *The Net and the Nation State*, 147–64.

core component of the *Lotus* decision—is not necessarily a natural leading light online.

I have here argued that we now—a bit more than 80 years after their publication—draw the curtain on the 'Harvard Draft' principles as the dominant point of departure for jurisdictional enquires. After all, given the complexity of jurisdictional considerations not least in today's society, the 'Harvard Draft' principles are unrealistically mechanical. They validate a dated and rigid focus on territoriality that makes impossible any progress in relation to the complexities associated with Internet jurisdiction. In fact, I have argued that our adherence to territoriality is, to a strong degree, based on a sleepwalking habit of uncritically accepting the territoriality principle as inevitable, rather than being founded on any particularly solid principles of international law.

Through what may be described as a deconstructive analysis, I have sought to demonstrate that, while the territoriality principle has a rather long, and relatively, consistent history—to the point of being seen as the jurisprudential core principle for jurisdiction—and while it runs like a *fil rouge* throughout both contemporary private international law and contemporary public international law, its initial attractiveness fades under the tension between its rigidity and the realities of our modern world.

I have pointed to how the distinction between, on the one hand, 'territorial' and, on the other hand, 'extraterritorial' jurisdictional claims is misleading and not merely stylistically odd. Further, I have provided several examples of how the law seeks to mitigate the destructive effect that a strict application of the territoriality principle may have. I have demonstrated that state practice has already moved beyond territoriality, territorial sovereignty, and the 'Harvard Draft' principles, and I have emphasised that territoriality is a component in a system that even on the most charitable assessment must be seen to be dysfunctional. Furthermore, there is, as stressed already, an increasing appreciation of the fact that, in a globalised world, state responsibilities do not end at the states' territorial borders. This is particularly clear in areas such as human rights law, antitrust law, environmental law, air law, and space law.

A key obstacle to conceptual and practical progress is found in the conflation between sovereignty and jurisdiction. In that context, we can usefully seek shelter in Orakhelashvili's clear and illuminating reasoning:

> Once we conceptually and analytically separate the issue of jurisdiction as an entitlement to exercise sovereign authority in relation to a particular person or matter, from the issue of actually carrying out that sovereign authority within the territory of another state, the dilemma with the extra-territoriality of state jurisdiction becomes far less acute. Similarly, breaking out of the influence of the territoriality cliché,

> creating analytical difficulties for the operation of more than one extra-territorial category of jurisdiction becomes far easier.
>
> Conflicts of one state's jurisdiction with another state's territorial supremacy are no longer so inescapable either. For jurisdiction as such does not consist in *exercising authority* of the state abroad, nor *securing obedience* abroad to the same authority. It merely manifests the willingness of the state to exercise its own authority within its own sovereign realm, but in relation to certain acts and events that take place abroad.[138]

Nevertheless, for some readers, the very idea of contemplating a departure from our paradigmatic obsession with territoriality will no doubt be startling. However, I am hopeful that most readers will, if they did not already do so, approach territoriality in a not quite so uncritical manner in the light of what has been presented in this chapter. In the next chapter, I outline what I see as an appropriate alternative to the current reliance on territoriality as the jurisprudential cornerstone of jurisdiction.

[138] Orakhelashvili (n 113) 11.

3

A New Jurisprudential Framework for Jurisdiction

For years the world has moved in the direction of increasing globalisation, which has guided much of the thinking in areas such as trade, defence, communication, the environment, and, indeed, the law. However, there are presently signs of globalisation losing its pace, and maybe also its 'halo'; international trade deals are rejected, cross-border cooperation are abandoned, and states are increasingly inward-looking and nationalistic. In this climate it is of the greatest importance that we have reasonably clear and well-defined, as well as reasonable and realistic, rules for jurisdiction, and the present exaggerated focus on territoriality does not allow for jurisdictional rules that are reasonable or realistic.

To move forward we must recognise that the territoriality principle and the associated but partially clashing concept of territorial sovereignty no longer serve as useful starting points for the analysis of jurisdictional claims. Not least for practical reasons, we need to search for a better point of departure for such tasks.

It is in the light of this world-view that I here have declared 'war' on the misguided assumption of territoriality as the jurisprudential cornerstone of jurisdiction; it is a just and legitimate war, and I hope to defeat the 'enemy' that is the widespread, unnecessary, and uncritical overreliance on territoriality.[1]

3.1 The Method: From Proxy Principles to Core Principles

Orakhelashvili has observed that '[f]or ascertaining the ambit of jurisdiction, territoriality is, at most, merely a conceptual starting-point, not the

[1] 'I have resolved never to start an unjust war, but never to end a legitimate one except by defeating my enemies', Karl XII <http://www.azquotes.com/quote/542917>, accessed May 2017.

Solving the Internet Jurisdiction Puzzle. First Edition. Dan Jerker B. Svantesson.

ultimate guide.'[2] In fact, I would go as far as to suggest that the territoriality principle and all the other 'Harvard Draft' principles are merely proxy principles for, and simplified expressions of, underlying core principles. They were after all constructed, through compromises, to reflect the legal practice at the time—they were a convenient summary of what courts did, what legislators thought, and so on. I hasten to acknowledge that the 'Harvard Draft' principles have largely been treated as jurisprudential core principles and therefore have adopted such a role on a practical level—they are the measuring stick against which jurisdictional claims are assessed. My claim is, however, that they should properly never have been viewed as such core principles—viewed thus they set inappropriately tight boundaries for our thinking in that we are unable to explore options that stray from the territoriality focus.

Searching the literature, it is clear that I am by no means the first to have sought to look behind principles, such as the territoriality principle, in search of their justification. For example, the 'Comment' to the 'Harvard Draft' points to Lewis's statement that '[t]he received rule as to the territoriality of criminal law rests on a sound basis. The territorial sovereign has the strongest interest, the greatest facilities, and the most powerful instruments for repressing crimes, whether committed by native-born subjects, or by domiciled aliens, in his territory.'[3] This important quote points essentially to three justifications for the territoriality principle: the territorial sovereign's (1) stronger interest, (2) greatest facilities, and (3) more powerful instruments, in its territory. These justifications—and indeed underlying focal points—for the territoriality principle have been lost along the way and, as applied today, the territoriality principle seems to need no justification; it is applied without thought and in the crudest way—a state has the exclusive right to regulate all that occurs in its territory for the simple reason that it occurs in its territory. In a similar way we may usefully consider what brought about a territorial focus in private international law. Looking at the Common Law system, for example, Nygh states that:

> [w]ith the territoriality of the law which they administered was coupled the inability of the common law courts to take notice of facts and customs occurring or prevailing outside the realm. The jury originally was called from the neighbourhood in which the facts in issue had occurred so that they, from their own knowledge, could

[2] Alexander Orakhelashvili, 'State Jurisdiction in International Law: Complexities of a Basic Concept' in Alexander Orakhelashvili (ed), *Research Handbook on Jurisdiction and Immunities in International Law* (Edward Elgar 2015) 9.

[3] Sir George Cornewall Lewis, *On Foreign Jurisdiction and the Extradition of Criminals* (John W Parker & Son 1859) 30, as referred to in 'Introductory Comment to the Harvard Draft Convention on Jurisdiction with Respect to Crime 1935' (1935) 29 Supp AJIL 443, 483.

determine the facts. As a result the venue had to be laid in the county where the facts in issue arose. At a time when the jury supplied both the evidence and the judgment of the facts, it was essential that trial of causes in crime or in tort be held in the *locus delicti*, just as issues concerning the making or construction of a contract should be tried in the *locus contractus* and issues concerning land or other assets in the *locus rei sitae*. In these circumstances the common law courts could not take cognisance of causes of action which had arisen out of the realm, since no jury could be summoned to try the facts.[4]

In other words, the focus on territoriality sprang from the necessity of having a jury with knowledge of the custom at the relevant *locus*. This also highlights that there are practical considerations that underpin the territoriality principle.

At any rate, whatever the status of the 'Harvard Draft' principles *de lege lata*, it is my firm view that they should not be relied upon as jurisprudential core principles *de lege ferenda*. Particularly when we are trying to apply the law to novel phenomena that need to become the subject of clear legal rules, we need to cut away the undergrowth of proxy principles such as those articulated in the 'Harvard Draft' and identify the core principles that are reflected in those proxy principles. Only then will we be able to focus on the considerations and values that truly are to be balanced.

Thus, in order to go forward, we must first take a few steps back and identify the jurisprudential core principles for which the 'Harvard Draft's' jurisdictional principles act as proxies—and that at any rate represent a better jurisprudential core for jurisdictional enquiries.

3.2 The New Paradigm

It would seem inappropriate to so strongly criticise the established thinking on jurisdiction without providing at least a sketch of what a new paradigm might look like. I will do so here. And while I discuss this new framework as consisting of core principles that always—at least in theory—have underpinned what I see as proxy principles such as the territoriality principle, the acceptance of the utility and legitimacy of the framework I advocate is, of course, not contingent on an acceptance of my 'core principles' versus 'proxy principles' claim.

[4] PE Nygh, 'The Territorial Origin of English Private International Law' (1964) 2 U Tas L Rev 28, 29–30.

At any rate, it seems to me that the essence of the jurisdictional principles expressed in the 'Harvard Draft' may be distilled into two core principles; that is, jurisdiction may only be exercised where (1) there is a substantial connection between the matter and the state seeking to exercise jurisdiction and (2) the state seeking to exercise jurisdiction has a legitimate interest in the matter. In other words, the 'Harvard Draft' principles seek to rely on mechanical binary proxies such as territory and nationality as shortcuts avoiding the real issues of substantial connection and legitimate interest—issues that always will be matters of degree rather than being binary.

I recognise that to say that 'legitimate interest' and 'substantial connection' are essential features of any acceptable jurisdictional claim may appear both truistic and vague. Yet these features cannot be omitted in any faithful account of the jurisprudential basis for jurisdiction nor can they be made more precise at the level of core principles.

Critics will no doubt question how my proposal relates to the (in some ways more complex) jurisdictional principles: that is, the protective principle and the universality principle. My answer to them is that the protective principle is, in fact, the clearest expression of a focus on substantial connections and legitimate interests; as to the universality principle, I think it appropriate to take the view that all states have a substantial connection to, and a legitimate interest in, the type of matters to which that principle relates.

In addition to the two core principles identified above, I would like to advance a third—the balance between the state's legitimate interests and other interests. While this principle is not as clearly part of the 'Harvard Draft' principles as are the other two, it can nevertheless be seen in the application of, for example, the protective principle.[5] Furthermore, as noted by Crawford: '[S]ufficiency of grounds for jurisdiction is normally considered relative to the rights of other states.'[6] Thus, while my proposed third principle goes further than what Crawford speaks of, one can nevertheless see hints of the thinking behind this third principle in current practice. Finally, and perhaps most importantly, adding the third principle to the other two may allow us to use the set of core principles to replace the other pieces of the puzzle also: that is, the doctrine of comity, territorial sovereignty, due diligence, and the duty of non-intervention as they apply in the jurisdictional context. Put bluntly, this third principle completes the picture.

[5] Noah Bialostozky, 'Extraterritoriality and National Security: Protective Jurisdiction as a Circumstance Precluding Wrongfulness' (2014) 52 Colum J Transnatl L 617, 621, noting that 'protective jurisdiction is presently formulated by reference to comparative state interests'.

[6] James Crawford, *Brownlie's Principles of Public International Law* (8th edn, OUP 2012) 457.

Summarising, the proposed jurisprudential framework for jurisdiction consisting of three core principles:

> In the absence of an obligation under international law to exercise jurisdiction, a state may only exercise jurisdiction where:
> (1) there is a substantial connection between the matter and the state seeking to exercise jurisdiction;
> (2) the state seeking to exercise jurisdiction has a legitimate interest in the matter; and
> (3) the exercise of jurisdiction is reasonable given the balance between the state's legitimate interests and other interests.

It is this jurisprudential framework I envisage as the new paradigm for jurisdiction; these are the principles I see as the core principles of jurisdiction.

This framework was originally unveiled in a special issue of the *American Journal of International Law Unbound*.[7] Two of the three experts providing comments on my proposal in that special issue devoted much energy to explaining why my proposal invokes feelings of *déjà vu* (Muir Watt[8]) and perhaps amount to 'old wine in new bottles' (Ryngaert[9]). However, that this is so is both obvious and intentional.

Few, if any, sensible legal improvements appear out of thin air, and my proposal is a result of extensive research. I have sought to anchor the proposal in established thinking without tying it directly to concepts overly burdened by their history. The three principles I advance are indeed not new inventions; they are 'old wine'. But they are old wine of the greatest pedigree—the type of old wine that is mature, desirably and beneficially old, and their *déjà vu*-like familiarity ought to be a welcomed and positive feature. And the 'new bottles' are more than fancy packaging. As acknowledged by Ryngaert:

> [W]hat is new is that he [I] *upgrades* these criteria [connection/interest/reasonableness] to *first-order* jurisdictional criteria. [...] Put differently, in the classic view, connection/interest/reasonableness only come into play once an assertion can be preliminarily be [*sic*] justified on the basis of a traditional permissive principle, in particular territoriality. In Svantesson's approach, we skip the first step and directly apply his core principles of connection/interest/reasonableness[.][10]

[7] Dan Svantesson, 'A New Jurisprudential Framework for Jurisdiction: Beyond the Harvard Draft' (2015) 109 AJIL Unbound 69 <https://www.asil.org/blogs/new-jurisprudential-framework-jurisdiction-beyond-harvard-draft>, accessed May 2017.

[8] Horatia Muir Watt, 'A Private (International) Law Perspective: Comment on "A New Jurisprudential Framework for Jurisdiction"' (2015) 109 AJIL Unbound 75, 109.

[9] Cedric Ryngaert, 'An Urgent Suggestion to Pour Old Wine Into New Bottles: Comment on "A New Jurisprudential Framework for Jurisdiction"' (2015) 109 AJIL Unbound 81.

[10] ibid 82.

I am indebted to Ryngaert for this elegant description of wherein the novelty lies, and on the whole, I am prepared to defend my proposed jurisprudential framework as something novel. Put in the fewest of words, the jurisprudential framework I advance has neither been presented in this form, nor as fulfilling this role, previously. However, what matters is not the extent to which my proposal is novel. What really matters is whether or not it can improve the law.

Finally, at risk of being accused of belabouring the obvious, the three core principles I advance are subject to the same meta-principles that apply to all laws. For example, where a law is kept secret, so as to make compliance with the law impossible, its moral legitimacy, or at least practical validity and usefulness, may be called into question. Thus, a jurisdictional claim that is supported by the three core principles I propose may nevertheless be objectionable where it falls foul of one of the meta-principles applicable to all law, such as that the law giving a court the right to exercise jurisdiction has been kept secret. The question of precisely which principles ought to be seen as such meta-principles need not detain us here.[11] It is quite simply a matter best discussed in a broader context as those principles are not specific to our context, and consequently, these paragraphs are not intended as a contribution to that debate.

3.2.1 Substantial Connection

Discussing territorial jurisdiction in relation to the Internet, Ryngaert states that:

> [i]n practice, States require a *substantial* territorial connection in order to establish jurisdiction. Such a connection may consist of (a) a production of effects within the territory, e.g. as a result of the alleged perpetrator purposely addressing content to territorially based users, or in the case of serious offenses, mere accessibility of a website (a variation on the objective territoriality principle, which comes, however, close to universal jurisdiction); (b) the location of the offender's computer within the territory via an identified IP address (a variation on the subjective territoriality principle); or (c) the storage of content on a local server.[12]

[11] Here it suffices to note that Fuller's eight routes to disaster in lawmaking constitute a useful starting point for a discussion of such meta-principles (Lon L Fuller, *The Morality of Law* (2nd edn, Yale University Press 1969) 39).

[12] Cedric Ryngaert, *Jurisdiction in International Law* (2nd edn, OUP 2015) 79–80.

While examples certainly can be found of such a *substantial* territorial connection being required, it may not be quite as widespread as one may think reading Ryngaert's statement. Furthermore, Ryngaert seems to suggest that the mere fact that content is stored on a server in a particular state amounts to a *substantial* territorial connection. However, focusing the jurisdictional enquiry on the location of the server has been criticised for some time now and, not least given how cloud computing operates, few would agree that content being stored on a server in a particular state amounts to a *substantial* territorial connection.

At any rate, the substantial connection requirement is well established in both public and private international law. For example, as far as private international law is concerned, we can point to the central role 'substantial connection' plays in Canada.[13]

Turning to public international law, in the Australian High Court's decision in the *Gutnick* case, Kirby J referred to 'the ordinary principle of public international law obliging a substantial and bona fide connection between the subject matter of a dispute and the source of jurisdiction of a national court over its resolution'.[14] The most well-known authorities discussing a 'connection requirement' are *Nottebohm*[15] and *Barcelona Traction*.[16] Indeed, Crawford asserts that '[i]f there is a cardinal principle emerging, it is that of genuine connection between the subject-matter of jurisdiction and the territorial base or reasonable interests of the state in question.'[17] While Crawford does not seek to anchor this thinking in the 'Harvard Draft' principles, and while he also points out that the territoriality principle is universally recognised,[18] his observation can perhaps nevertheless be seen as supportive of my first two principles.

[13] See eg s 10(i) of the *Court Jurisdiction and Proceedings Transfer Act* (*CJPTA*), as well as caselaw such as: *Society of Composers, Authors and Music Publishers of Canada v Canadian Assn of Internet Providers* 2004 SCC 427, [2004] 2 SCR 427 [54]–[63] and *Club Resorts Ltd v Van Breda* 2012 SCC 17.

[14] *Dow Jones and Company Inc v Gutnick* (2002) 210 CLR 575 [622].

[15] *Nottebohm (Liechtenstein v Guatemala)* ICJ Rep 4 et seq (1955).

[16] *Barcelona Traction (Belgium v Spain)* ICJ Rep 42 (1970). See further Ryngaert (n 13) 156–7.

[17] Crawford (n 6) 457. See also Bruno Simma and Andreas Th Müller, 'Exercise and Limits of Jurisdiction' in James Crawford and Martti Koskenniemi (eds), *The Cambridge Companion to International Law* (CUP 2012) 149, stating that '[t]he international law of jurisdiction thus seems to perform a comprehensive evaluation and weighting of factors that add up to, or detract from, the extent of a state's affectedness that is needed for that state to lawfully claim jurisdiction. This constitutes the very essence of the "substantial link" doctrine.'

[18] Crawford (n 6) 458.

3.2.2 Legitimate Interest

An overlap between the 'substantial connection' and the 'legitimate interest' requirements—in the sense of where one is present, the other is likely to be present too—ought to be relatively common. However, it would be a mistake to conclude, based on this, that the two tests are the same, and a distinction is commonly drawn between the two.[19] They are conceptually distinct; and after all, it is not difficult to imagine situations where the two requirements do not overlap. Thus, emphasising that a substantial connection does not automatically give rise to a legitimate interest is an important feature of the proposed framework.

At any rate, it is commonplace to point to the significance of legitimate interests for the evaluation of jurisdictional claims. Indeed, relying on the work of Oehler, Ryngaert argues that we can trace the legitimate interest focus all the way back to Bartolus and thirteenth-century Italy.[20]

Further, and of even greater significance, I am by no means the first to argue that when we look behind the 'Harvard Draft' principles, we find legitimate interest as an underlying core principle. Indeed, some commentators focus seemingly exclusively on legitimate interest:

> [T]he customary international law of criminal jurisdiction is based on a perceptible, if somewhat ill-defined, set of principles regarding the legitimate prosecutorial interests of states. In most criminal cases, those underlying principles are not relied upon explicitly, because the customary law of criminal jurisdiction recognizes an identifiable set of valid bases for jurisdiction. When jurisdiction can be justified by reference to one of those recognized bases, as is usually the case, reiteration of the underlying principles concerning the legitimate prosecutorial interests of states is not necessary. The usual list of internationally recognized bases of jurisdiction includes: territoriality, nationality, protective principle, universality, and passive personality (the last being the least robustly accepted). Some influential authorities maintain that this list is exclusive. Because the list of recognized bases for jurisdiction is not arbitrary but has developed to reflect an evolving delineation of the legitimate prosecutorial interests of states, it is unlikely that the list is actually closed. Rather, what appears to be true is that jurisdictional bases that are already recognized are uncontroversially acceptable, while the legitimacy of claimed new bases must be determined.[21]

The conclusion we can draw from the above is that legitimate interest is a well-established central focal point for jurisdiction.

[19] See eg Ryngaert (n 12) 39. [20] ibid.

[21] Madeline Morris, 'High Crimes and Misconceptions: The ICC and Non-Party States' (2001) 64 L & Contemp Probs 13, 47–48 (footnote omitted).

3.2.3 Interest Balancing

Interest balancing is commonplace in law, and not least in the international law arena. For example, as already noted, Crawford makes the point that the 'sufficiency of grounds for jurisdiction is normally considered relative to the rights of other states'.[22] Similarly, relying on a wealth of sources, Ryngaert asserts that 'even if a link or interest could be discerned, and a State could rely on an accepted permissive principle, the legality of the exercise of jurisdiction under international law might depend on the harm that such exercise causes to other sovereigns'.[23] This also points to a balancing exercise of the nature envisaged in my proposed framework. And discussing why the Common Law shows restraint when it comes to jurisdictional claims in private international law, Keyes refers to two grounds: 'The first is the forum state's concern not to offend the sovereignty of other countries. [...] The second reason for the courts' reluctance to assert jurisdiction extraterritorially was a concern to spare foreign defendants the burden in costs and inconvenience of defending themselves in a distant forum.'[24] Both these grounds fit well within the interest balancing I envisage for the third principle. Support for the type of thinking represented in the third principle can also be found in, for example, the public international law 'principle of proportionality'[25] and private international law's *forum non conveniens* doctrine.[26] But perhaps even stronger support is found in the principle of reasonableness[27] articulated in §403(2) and (3) of the *Restatement (Third) of US Foreign Relations Law* (1987). As is well known, §403(2) outlines a list of criteria to be used in assessing the reasonableness of exercising jurisdiction:

(2) Whether exercise of jurisdiction over a person or activity is unreasonable is determined by evaluating all relevant factors, including, where appropriate:

[22] Crawford (n 6) 457. [23] Ryngaert (n 12) 39.

[24] Mary Keyes, *Jurisdiction in International Litigation* (Federation Press 2005) 51. In our context, it is worth observing that the convenience argument may be weaker today than it once was. As noted by the Court in *Hyde v Agar*: 'Contemporary developments in communication and transport make the degree of 'inconvenience and annoyance' of a qualitatively different order to that which existed in 1885. It follows that a modern court applying the Rules must be sensitive to the actual degree of 'inconvenience and annoyance' which will be imposed in the circumstances of a particular case, rather than assuming that any person or resident abroad must suffer such consequences in substantial degree' (*Hyde v Agar* (1998) 45 NSWLR 487, at 507) This does not, however, undermine the idea that the interests of the parties needs to be part of the calculation.

[25] See further Ryngaert (n 12) 158–60.

[26] Dan Svantesson, *Private International Law and the Internet* (3rd edn, Kluwer Law International 2016) 484–98.

[27] Ryngaert (n 12) places great emphasis on this principle in his work, *Jurisdiction in International Law*, 158–60.

(a) the link of the activity to the territory of the regulating state, ie, the extent to which the activity takes place within the territory, or has substantial, direct, and foreseeable effect upon or in the territory;
(b) the connections, such as nationality, residence, or economic activity, between the regulating state and the person principally responsible for the activity to be regulated, or between that state and those whom the regulation is designed to protect;
(c) the character of the activity to be regulated, the importance of regulation to the regulating state, the extent to which other states regulate such activities, and the degree to which the desirability of such regulation is generally accepted;
(d) the existence of justified expectations that might be protected or hurt by the regulation;
(e) the importance of the regulation to the international political, legal, or economic system;
(f) the extent to which the regulation is consistent with the traditions of the international system;
(g) the extent to which another state may have an interest in regulating the activity; and
(h) the likelihood of conflict with regulation by another state.

To this §403(3) adds:

> When it would not be unreasonable for each of two states to exercise jurisdiction over a person or activity, but the prescriptions by the two states are in conflict, each state has an obligation to evaluate its own as well as the other state's interest in exercising jurisdiction, in light of all the relevant factors, Subsection (2); a state should defer to the other state if that state's interest is clearly greater.

This structure has been criticised. For example, Meessen expresses the view that '[a]s long as the world continues to be composed of more than one state, there will be no agreed understanding on what is reasonable in the meaning ascribed to that term in section 403(2).'[28] This scepticism, not to say pessimism, is perhaps justified. But it does not detract from my suggestion that the balancing I propose as my third principle, as such, is an accepted component of the jurisdictional inquiry.

I am not suggesting that the balancing occurring under my third principle necessarily needs to be guided by *all* of what is proposed in §403(2) of the *Restatement (Third) of US Foreign Relations Law*. And I am not suggesting that the balancing occurring under my third principle necessarily needs to be guided *only* by the criteria outlined in §403(2). However, it is clear that

[28] Karl M Meessen, 'Conflicts of Jurisdiction under the New Restatement' (Summer 1987) 50 L & Contemp Probs 47, 59.

aspects of §403(2) and (3) of the *Restatement (Third) of US Foreign Relations Law* may be useful in the context of my third principle, and lends support to its inclusion in a jurisprudential framework for jurisdiction.[29]

On a more general note, it has been observed that 'the Restatement Third places less emphasis on territoriality. Instead, it focuses on elements such as the nature of the activity, the effect on the regulating state, and the interests of both the regulating and territorial states, which interests are measured by the amount of regulation generally exercised.'[30] The type of 'interest balancing' I advocate has a history in some other areas of law, such as competition/antitrust law,[31] and there are signs of an increasing appetite for interest balancing also in other fields.[32] Perhaps the area of law that can provide the most guidance for the application of balancing to be undertaken within the third core principle is that of 'comity'.[33] Drawing upon the work of Maier, Johnson and Post note that comity 'functions as a constraint on the strict application of territorial principles that attempt to reconcile "the principle of absolute territorial sovereignty [with] the fact that intercourse between nations often demand[s] the recognition of one sovereign's lawmaking acts in the forum of another" '.[34]

Indeed, some commentators may be inclined to suggest that what the third core principle aims at is just the same as what the concept of comity is aiming

[29] For an interesting discussion of §403(2) and (3) of the *Restatement (Third) of US Foreign Relations Law*, see eg Ryngaert (n 13) 162–71.

[30] Kathleen Hixson, 'Extraterritorial Jurisdiction under the Third Restatement of Foreign Relations Law of the United States' (1988) 12 Fordham Intl LJ 127, 137.

[31] In the context of the interest-balancing test, Ginsburg has noted that '[o]ne general response in public law has been proportionality, which began as a principle of administrative law in 19th century Prussia, but has spread rapidly to become the adjudicative technique *par excellence* for constitutional decision-makers the world over. Proportionality is also a principle of international law, applied in contexts as diverse as the WTO, the International Covenant of Civil and Political Rights, and international humanitarian law. It provides a rigorous framework for evaluating competing interests under high-stakes conditions' (Tom Ginsburg, 'Comment on Svantesson, A New Jurisprudential Framework for Jurisdiction.' (2015) 109 AJIL Unbound 86, 87).

[32] See eg B Van Alsenoy and M Koekkoek, 'Internet and Jurisdiction after Google Spain: The Extraterritorial Reach of the "Right to be Delisted" ' (2015) 5 Intl Data Priv L 105–20, 117–18. Admittedly, Van Alsenoy and Koekkoek present their balancing exercise within a framework focused on connecting factors in relation to the *territory* of the forum state.

[33] For in-depth discussions of comity, refer to interesting research work carried out by Schultz, such as: Thomas Schultz and David Holloway, *The Principle of Comity in Private and Public International Law* (CUP 2017), and Thomas Schultz and Jason Mitchenson, 'Navigating Sovereignty and Transnational Commercial Law: The Use of Comity by Australian Courts' (2016) 12(2) J Priv Intl L 344–78 <http://dx.doi.org/10.1080/17441048.2016.1206704>, accessed May 2017.

[34] David Johnson and David Post, 'Law and Borders—The Rise of Law in Cyberspace' (1996) 48 Stan L Rev 1367, 1392, referring to Harold G Maier, 'Remarks' (1990) 84 Proc Am Soc Intl L 339, 339 (amendment as per original).

for. However, as has already been emphasised, the concept of comity lacks a uniform interpretation to the degree that, while some see it as of the greatest importance, others view it as useless. This 'baggage' disqualifies the option of structuring the third core principle around comity as such. Nevertheless, the lessons learnt from the application of the doctrine of comity ought to be most useful in the context of the balancing to be undertaken within the third core principle.

At any rate, I have intentionally kept the wording of the third principle vague in that I make no strict commitment as to what type of competing interests exclusively may be considered. In fact, it would be unwise to do so in the abstract. Under the current paradigm, it could be suggested that those interests are limited to competing *state* interests. However, such a focus is clearly inappropriately restrictive. There are strong and obvious reasons to broaden the scope so as also to take account of, for example, the interests of the party or parties being subjected to the jurisdictional claim,[35] as well as the interests of the broader international community.[36] This is a question that deserves to be discussed in detail. However, I shall not, for the moment, explore that topic any further, as it will find a more congenial environment in the text in this chapter, and I will return to it then.

Before examining the role that the framework already proposed may have, one additional observation must be made. As Mills points out there is a

[35] See eg the interesting work of Ireland-Piper regarding whether the 'abuse of rights' doctrine might be helpful in seeking to maintain an appropriate balance between the rights of states and of individuals. Danielle Ireland-Piper, 'Prosecutions of Extraterritorial Criminal Conduct and the Abuse of Rights Doctrine' (2013) 9 Utrecht L Rev <http://www.utrechtlawreview.org/index.php/ulr/article/download/243/237>, accessed 3 April 2017, and Danielle Ireland-Piper, *Accountability in Extraterritoriality* (Edward Elgar 2017), in which Ireland-Piper notes that 'the principles of jurisdiction under international law do not adequately resolve competing claims to jurisdiction and are primarily concerned with the relationship between States and not as between the State and the individual' (2). See also Ryngaert (n 12) 60–1.

[36] Discussing the focus on contacts and interest in §403 of the *Restatement (Third) of US Foreign Relations Law*, Ryngaert (n 12) 190 has noted that 'the method of § 403 may fail to do justice to the genuine interests of the international community, as it may unduly lay emphasis on *State* contacts and interests. A balancing of interests then risks resulting in a jurisdictional outcome that is harmful to the international community. A State may have the strongest connections to, or the strongest interests in, regulating a situation but decide not to regulate. The absence of regulatory intervention may serve the interests of that State, but produce adverse effects at a global level'. However, this criticism cannot contaminate my model, since what I am proposing is not multilateralism's search for the one and only strongest/closest connection or interest in the matter. Rather, the framework I am advocating falls within unilateralism and asks: (1) whether the state's legitimate interest in the matter is sufficiently strong, (2) whether the state's substantial connection to the matter is sufficiently strong, and (3) whether the exercise of jurisdiction is reasonable given the balance between the state's legitimate interests and other interests.

growing recognition that in some circumstances the exercise of national jurisdiction may, under international law, be a question of duty or obligation rather than right or discretion. To put this another way, the regulation of jurisdiction in international law needs to be reconceived as not merely a 'ceiling', defining the maximum limits of state power, but also (in some contexts) as a 'floor', reflecting minimum requirements for the exercise of regulatory power by states in order to satisfy their international obligations.[37]

The recognition of such duties and obligations is, in fact, found already in the 'Harvard Draft': 'States may be under an obligation to exercise penal jurisdiction in certain cases by virtue of principles of customary international law or international agreement.'[38] In the light of this, any framework for jurisdiction must necessarily take account of such duties and obligations, and it should be noted that an obligation under international law to exercise jurisdiction may stem both from specific treaty undertakings[39] and from customary international law, for example in relation to access to justice.[40]

Due to the inclusion of the interest balancing in the third principle, the framework I advance works both in relation to the exercise of jurisdiction as a right and to the exercise of jurisdiction as an obligation. For example, where there is a substantial connection between the matter and a state and that state has a legitimate interest in the matter, that state may be under an obligation to exercise jurisdiction based on the balance between the state's legitimate interests and other interests, such as the interests of a party that will be denied access to justice unless that state exercises jurisdiction. Thus, the framework does indeed both provide a 'floor' and a 'ceiling'.

Moving on, a special mention is in order about just how strong the interest of legal or natural persons may be as a factor taken into account in the balancing in the third component of the proposed framework. Mills has pointed to how the role of party autonomy has expanded, particularly in private

[37] Alex Mills, 'Rethinking Jurisdiction in International Law' (2014) 84 Brit YB Intl L 209 (internal footnote omitted).

[38] 'Introductory Comment to the Harvard Draft Convention on Jurisdiction with Respect to Crime 1935' (1935) 29 Supp AJIL 443, 446.

[39] Mills (n 37) 210 refers to Art 7(1) of the Convention Against Torture as an example of this: 'The State Party in the territory under whose jurisdiction a person alleged to have committed any offence referred to in article 4 is found shall in the cases contemplated in article 5, if it does not extradite him, submit the case to its competent authorities for the purpose of prosecution'.

[40] Mills discusses this at length. Here it suffices to note his statement that: '[i]t has long been recognised that states owe obligations to meet a "minimum standard of treatment" in respect of their dealings with each other's nationals. The standard of treatment includes a requirement for states to afford "adequate judicial protection and effective legal remedies for repairing invasions of rights" for foreigners, whether natural or legal persons, typically through access to domestic courts' (ibid 214 (internal footnote omitted)).

international law, but in his view (and I agree), also in public international law. This has implications for the traditional perspective of jurisdiction: 'If a state exercises jurisdiction or applies its law in civil proceedings based purely on consent by the parties, this is difficult to reconcile with the traditional public international law requirement that jurisdiction must be justified by a substantial objective connection, typically territoriality or nationality.'[41] Under the core principles advanced here, party-autonomy—in the sense that a state may exercise jurisdiction and/or apply its law in civil proceedings based purely on consent by the parties—is primarily accommodated through the interest balancing that occurs in the third step. However, this, of course, is not enough to ensure the framework survives the challenge posed by jurisdiction based on party-autonomy. After all, if the parties can choose a jurisdiction without any objective connection or interest in their relationship, that could be seen to render the first two principles optional.

This is a most serious matter, and I have only been able to think of two possible solutions. We could say that jurisdiction based on party-autonomy should only be upheld where the state of the court has a substantial connection to, and legitimate interest in, the matter. Such an approach would best serve the integrity of the jurisprudential framework I advance. However, while a connection is required in some situations in some legal systems,[42] incorporating such a requirement on the jurisprudential level discussed here is simply unrealistic due to the harmful impact it would have upon party-autonomy.

The better view is that the parties' choice of a particular jurisdiction or applicable law, in itself, creates a sufficiently substantial connection to, and legitimate interest in, the matter for the state favoured by the parties. This approach, while admittedly somewhat inelegant, upholds the integrity of the jurisprudential framework without unduly interfering with party-autonomy.

I hasten to add, however, that this does not by any means indicate that courts are forced to uphold the parties' choice of jurisdiction or applicable law. As part of the third step, a court may well conclude that the balance favours not upholding the parties' choice.[43] Constructing detailed second-level rules as to when a court is to uphold the parties' choice will be a task for each state's legislator or the drafters of international instruments.

[41] ibid 233.

[42] eg some courts only allow party-autonomy-based jurisdiction where the chosen court has a 'practical connection with the dispute' (*Civil Procedure Law of the People's Republic of China* (2008), Art 242).

[43] For a discussion of some reasons why a court may reach such a conclusion, see eg Svantesson (n 26) 429–31.

3.3 The Role of the New Framework

The paradigm shift I advocate, from the territoriality principle and the other 'Harvard Draft' (proxy) principles to the proposed core principles, represents an important philosophical and theoretical change. Its main aim is to change the point of departure when discussing jurisdictional claims, thereby altering the jurisprudential underpinnings on which we base our understanding of jurisdictional claims and their legitimacy, or lack of legitimacy. In a sense, at the minimum, the framework can be seen as more of a tool to guide our thinking on jurisdiction than as strict legal norms.

It should also be emphasised that the shift would see no practical change in non-controversial areas of jurisdiction. For example, a state would obviously have a substantial connection to, and a legitimate interest in, a traffic offence occurring within its territory, and the principle of balancing between that state's legitimate interests and other competing interests ought not to cause any complications in such matters. The absolute majority of cases, both offline and online, will involve a similarly natural connection between territoriality, on the one hand, and substantial connection and a legitimate interest on the other hand. This is important as it means that the proposed change actually is rather non-threatening also for those who treasure the territoriality principle. After all, the territoriality principle is neither necessarily, nor always, incompatible with the proposed core principles. Thus, in any setting where the territoriality principle works well, territoriality can remain in place. The only change in such situations is that the territoriality principle is applied as a second-level rule rather than as a core principle. On a practical level this does not make much difference, although it may mean that it is easier to introduce exceptions to the territoriality-focused rule where such exceptions are needed.

At the same time, despite the theoretical aim, and even though the new core principles do not preclude the use of territoriality as a second-level rule, the paradigm shift also provides benefits on the practical level. Done carefully and diligently, the proposed paradigm shift would see us being much better equipped to address what are now controversial and problematic areas. It will allow us to think more creatively rather than just being mechanically binary. It would, for example, free us from the thinking that State A must always have a possible jurisdictional claim over all aspects of data that happened to be located on a server located in State A. For example, in the mentioned *Microsoft Warrant* case,[44] we would not simply ask the (impossible)

[44] *In the Matter of a Warrant to Search a Certain E-mail Account Controlled and Maintained by Microsoft Corporation*, No 14-2985, 2016 WL 3770056 (2d Cir July 14, 2016).

question of whether the United States law enforcement activity would take place on Irish territory or not. We would look more broadly at the connections and interests involved. The significance of this become clear where we place the *Microsoft Warrant* case side by side with the less famous *Google Warrant* case.[45]

Both cases involve law enforcement agencies seeking user data and relying on a warrant pursuant to §2703 of the *Stored Communications Act*, 18 USC §§2701 et seq. In both cases, the company handed over data known to be in the US. And in both cases, the law enforcement agencies wanted more.

All this is important, but then so are the differences. All account holders in the *Google Warrant* case reside in the US, while it seems that the relevant account holder in the *Microsoft Warrant* case was an Irish citizen residing in Ireland.[46] Further, in the *Google Warrant* case the crimes the account holders were suspected of committing occurred solely in the US, while this is not so in the *Microsoft Warrant* case. Arguably of even greater significance, it was established that the requested data in the *Microsoft Warrant* case was located on servers in Ireland. In contrast, given that 'Google stores data in a dynamic network which distributes the data, sometimes in bits and pieces, to servers located domestically and in foreign countries',[47] Google contended that 'it does not currently have the capability, for all of its services, to determine the location of the data'.[48] One important implication of this is that, while the relevant law enforcement agency in the *Microsoft Warrant* case could have gained access to the data under the Mutual Legal Assistance Treaty (MLAT) structure, no such avenue existed in the *Google Warrant* case scenario; after all, if you do not know to which state you should direct the MLAT request, you cannot send any such request at all.[49] The availability of alternative avenues is of obvious relevance. However, the Court also saw Google's structure as relevant in relation to another matter. Discussing risks to international comity, the Court noted:

> Even if the interference with a foreign state's sovereignty is implicated, the fluid nature of Google's cloud technology makes it uncertain which foreign country's sovereignty would be implicated when Google accesses the content of communications in order to produce it in response to legal process. As explained earlier, Google's

[45] *In re Search Warrants* Nos 16-960-M-01and 16-1061-M to Google.

[46] Aaron Rogan, 'US sues Microsoft for Silk Road suspect's email' (*The Times* [online], 22 June 2016) <https://www.thetimes.co.uk/article/us-sues-microsoft-for-silk-road-suspects-email-snq0tlqj0>, accessed 8 April 2017.

[47] *In re Search Warrants* Nos 16-960-M-01and 16-1061-M to Google, 7.

[48] ibid 7.

[49] See further, ibid 27–29.

architecture not only divides user data among data centers located in different countries, but also partitions user data into shards. Furthermore, the data automatically moves data from one location on Google's network to another as frequently as needed, to optimize for performance, reliability and other efficiencies. [. . .] Because of the structure of this system, Google cannot say with any certainty which foreign country's sovereignty would be implicated when Google accesses the content of communications in order to produce it in response to legal process. Before a court bars the Government from using a judicially approved search warrant to require disclosure of user data that constitute evidence of crimes, it would do well not to be controlled by possibilities and legal abstractions, but to focus instead on realities.[50]

It is, of course, important not to confuse the matter of a lacking knowledge of *which* state's sovereignty is being interfered with, on the one hand, and the matter of whether *any* state's sovereignty is being interfered with on the other hand. And the Court's reasoning here may be accused of failing to recognise this distinction. Nevertheless, there is no doubt some merit in the issue to which the Court seeks to bring our attention.

As the score stands at the time of writing, Microsoft has been successful in the latest decision in its dispute and Google has been unsuccessful in the latest decision in its dispute, but both matters may be appealed. I am not here seeking to comment on the outcomes of the cases as such. Rather, I wish to use the similarities and differences in the background facts of these cases to illustrate a serious failing of the territoriality principle and how the framework I have advanced is better equipped to deal with matters such as these.

Under territoriality thinking, we may, for example, simply ask whether the data is stored in the US or not. If it is, then there is no jurisdictional problem, and if it is not, it is beyond the reach of the relevant law enforcement agency. Nice, simple, and outrageously inadequate. In adopting such a territoriality-focused single-factor test, we are forced to overlook matters such as: (1) the location, nationality, and residence of the suspects; (2) the situs of the crime; (3) the availability of alternative means to access the data; and (4) the degree—if any—of interference with the sovereignty of other states. Treating these matters as irrelevant is seriously irresponsible and points to a fatal flaw in the adopting of a territoriality-focused single-factor test. Indeed, this shows the inescapable flaw in *all* single-factor tests applied in this setting. After all, the comparison between the *Microsoft Warrant* case and the *Google Warrant* case has identified several factors that must affect our analysis of the circumstances under which law enforcement agencies in one state can access data. In the light of this, it does not matter how carefully we select

[50] ibid 25–26.

which factor to focus on; the problem does not only lie in which factor we select, rather it is found in the fact that we then ignore other factors that also must be considered.

Under the jurisprudential framework for jurisdiction I advanced, a factor such as whether the crime is domestic or not will matter for whether there is legitimate interest. The nationality and residence of the suspect will impact the assessment of the requirement of a substantial connection. And, the degree—if any—of interference with the sovereignty of other states would be an important factor in the balancing occurring under the third principle.

Anyhow, in light of all the above, it may be that the usefulness of the proposed paradigm shift will be evident in the odd 5 per cent of controversial situations and not in the standard 95 per cent of situations. Despite this, it would be a serious mistake to think that the paradigm shift does not fulfil an important function, just because its practical benefits are seen in exceptional rather than in standard cases. First of all, it is indeed the exceptional situations, not the standard situations, that truly put our legal norms to the test and bring attention to needs for reform. And second, one has a duty to aim for as useful legal norms as possible, and where an improvement can be made it would truly be a sin to neglect improving the law just because the standard situations are already catered for under a current, flawed, legal structure.

In addition, the jurisprudential framework I advance is—as is discussed throughout—also useful as a tool to explain, or address, several theoretical conundrums. For example, the framework can be used to take a position on a question that has preoccupied international lawyers at least since the *Lotus* decision; that is, in the words of Ryngaert:

> Under public international law, two approaches could logically be taken to the question of jurisdiction. Either one allows States to exercise jurisdiction as they see fit, unless there is a prohibitive rule to the contrary [the '*Lotus* principle'], or one prohibits States from exercising jurisdiction as they see fit, unless there is a permissive rule to the contrary.[51]

Neither of these approaches is ideal or even realistic, and it should be recalled that the principles derived from the *Lotus* case do indeed seem to be somewhat more nuanced than this description suggests. In addition, on my reading, the second approach seems entirely unrealistic in that it would mean that states have had no grounds to claim jurisdiction *at all* until the time at which public international law expressly recognised certain jurisdictional grounds.

[51] Ryngaert (n 12) 29.

Anyhow, applying my proposed framework to this matter, we may conclude that, where a state has a substantial connection to, and legitimate interest in, a matter, and where there are no competing interests, there is no need to search for any specific permissive rules of public international law. In contrast, where any of these circumstances are lacking, jurisdiction may only be claimed where there is a specific permissive second-level rule of public international law.

3.4 Pre-empting (Some of) the Potential Objections

The central role of cross-border communication must necessarily, in my view, be borne in mind when we discuss jurisdiction. In other words, I am admittedly and openly pro-communication across borders. The combination of being pro-communication and being 'territoriality-agnostic', or perhaps even a 'territoriality-nihilist', may seem to place me in the same camp as the Cyber-libertarians, such as Barlow[52] and Johnson and Post.[53] However, I can only subscribe to some of the ideals they famously promoted. Most importantly, I am not advocating treating the Internet as being beyond the law. In fact, my proposal is clearly and distinctly supportive of applying law to the online environment.

Nevertheless, commentators such as Goldsmith and Wu, who clamour for the supremacy of territoriality, would presumably still accuse me of failing 'to understand the many faces and facets of territorial governmental coercion'.[54] However, in response, should one be needed, I would suggest that they are failing to properly distinguish 'law', on the one hand, and 'power' on the other. As noted by Hurrell, 'All states and social orders need to gain the authority and legitimacy that the possession of crude power can never on its own secure. All major powers face the imperative of trying to turn a capacity for crude coercion into legitimate authority.'[55]

There is no doubt that location, and therefore territoriality, always will affect the physical possibility of enforcing one's will. Irish law enforcement may not necessarily be able to physically delete the data held by Microsoft

[52] JP Barlow, *A Declaration of the Independence of Cyberspace* (8 February 1996) <https://projects.eff.org/~barlow/Declaration-Final.html>, accessed 3 April 2017.

[53] Johnson and Post (n 34).

[54] Jack Goldsmith and Tim Wu, *Who Controls the Internet? Illusions of a Borderless World* (OUP 2008) 184.

[55] Andrew Hurrell, 'International Law 1989–2010: A Performance Appraisal' in James Crawford and Sarah Nouwen (eds), *Selected Proceedings of the European Society of International Law 2010* (3rd vol, Hart 2012) 10.

in Ireland that is requested by the United States government in the *Microsoft Warrant* case.[56] However, by blowing up the data centre in question they can certainly make the transfer of the data impossible assuming that the data in question exists only on servers in Ireland. So, yes, Goldsmith and Wu are correct to emphasise the relevance of territorial governmental coercion.[57]

However, and this is important, we are not obligated to base our laws exclusively on power. In fact, quite the opposite is true—it is important that law is more than a tool to give legitimacy to the strongest party; as Akehurst puts it: '[I]neffectiveness is not the same as illegality.'[58] Thus, there is nothing surprising about the fact that our laws may divide jurisdictional rights along different fault lines to those lines along which physical coercive power is delineated. In the light of this, Goldsmith and Wu's emphasis on the importance of how national governments use coercion to enforce national laws within their territories[59] does not represent a challenge to, or obstacle for, the territoriality-nihilistic jurisprudential framework for jurisdiction advocated here.

As is clear from the works referred to throughout, while what I advocate above marks a distinct departure from tradition, I am by no means alone in my call for a reassessment of the jurisdictional framework provided by the territoriality principle and the 'Harvard Draft'. And it is clear that one need not seek refuge in the largely discredited, but still interesting, writings of the early Cyber-libertarians[60] to find support for the notion that the dominant focus on territoriality is a poor fit with our contemporary world.

The discussion has already tackled some of the potential challenges posed to the jurisprudential framework I advance, the most serious being the difficulty of properly accommodating, and accounting for, jurisdiction based on party-autonomy. In this section I will, however, attempt to address some other potential objections that could be (and have been) raised against my proposal.

Commentators prone to panic-stricken exaggerations will no doubt see the proposed move away from territoriality as an attack on the very fundaments of the international system. After all, under public international law,

[56] *In the Matter of a Warrant to Search a Certain E-mail Account.*

[57] A particularly interesting example of the importance of geography in networks, from a physical enforcement perspective, is found in the work on so-called 'Alibi Routing': see further Dave Levin et al, 'Alibi Routing' (2015) Sigcomm <https://alibi.cs.umd.edu/papers/alibi_sigcomm15.pdf>, accessed 3 April 2017.

[58] Michael Akehurst, 'Jurisdiction in International Law' (1972–73) 46 BYIL 145, 181.

[59] Goldsmith and Wu (n 54) 184.

[60] For an introduction to the Cyber-libertarianism prevalent in the early days of WWW communications, see eg Dan Svantesson, 'Celebrating 20 years of WWW—a Reflection on the Concept of Jurisdiction' (2012) 6(1) Masaryk UJL & Tech 177, 179–83.

'territory is the denotation of a certain physical and geographical space that constitutes one of the qualifications for Statehood',[61] and thus, an attack on territoriality could be seen as an attack on statehood.[62] However, such a claim lacks sting.

It is important to distinguish between territoriality as, on the one hand, a component of sovereignty and statehood and, on the other hand, a foundation for jurisdiction. We may very well keep the first while moving away from the latter. In other words, we can move away from territoriality as a foundation for jurisdiction without necessarily abandoning territoriality as a component of sovereignty and statehood. If we recognised this, we actually have more options than might have been thought.

I am not suggesting that we abandon the concept of territoriality wholesale and in every possible setting. My proposal is considerably more modest than that. All I am suggesting is that we abandon territoriality as the jurisprudential core of our thinking on jurisdiction. We may well continue to look to 'territory' as an essential element of statehood, and allowing states to, 'at least for the time being, remain keystones—albeit re-positioned ones—of the global governance architecture'.[63] In other words, my proposal acknowledges the role of states and does not enter into the bigger question of an overhaul of the international legal system as such.

The modesty of my proposal led Muir Watt to conclude that 'the changes advocated seem unlikely to fulfil their ambitious promise of paradigm renewal'.[64] Indeed, in her contribution to the symposium issue of the *American Journal of International Law Unbound* that discusses my proposal, Muir Watt emphasised the importance of guarding 'against the trivialization of paradigm change (in law or elsewhere)'.[65] To this she added, 'In this respect, it is very doubtful that a change from territory as dogmatic principle to the open-ended formulation of connections and interests constitutes an epistemological upheaval of sufficient significance to qualify as a change in legal consciousness.'[66] Admittedly, I do not see any intrinsic value in attaching the term 'paradigm shift' to my proposal. However, it seems beyond

[61] Benedetta Ubertazzi, *Exclusive Jurisdiction in Intellectual Property* (Mohr Siebeck, 2012) 137.

[62] For a discussion of the role of 'territory' as a criterion for statehood, see further James Crawford, *The Creation of States in International Law* (2nd edn, OUP 2006) 46–52; Karen Knop and Susan Marks, 'The War against Cliché: Dispatches from the International Legal Front' in Christine Chinkin and Freya Baetens (eds), *Sovereignty, Statehood and State Responsibility: Essays in Honour of James Crawford* (CUP 2015) 96–116.

[63] Günther Handl, 'Extra-territoriality and Transnational Legal Authority' in Günther Handl, Joachim Zekoll, and Peer Zumbansen (eds), *Beyond Territoriality* (Martinus Nijhoff 2012) 6 (footnote omitted).

[64] Muir Watt (n 8) 75. [65] ibid 79. [66] ibid.

intelligent dispute that we, as far as jurisdiction is concerned, currently work in a territoriality and 'Harvard Draft'-centric paradigm. And it seems equally obvious that my proposal would take us away from that paradigm. Hence, while my proposal is limited in scope to the area of jurisdiction, it can reasonably be seen as a proposal for a paradigm shift within that field, even though it is not exactly the 'radical overhaul of the law of jurisdiction'[67] that Muir Watt perhaps would like to see.[68]

At any rate, even if it is accepted that the practical implications of the paradigm shift are restricted to exceptional cases, some may argue that the suggested paradigm shift would see us lose the simplicity that is provided by the mechanical binary proxy principles such as territory and nationality. And it is, of course, true that the core principles I advance are rather general in nature and thereby vague.[69] This is significant as Ginsburg points out in his contribution to the mentioned symposium issue of the *American Journal of International Law Unbound*. Ginsburg agrees with my proposal but emphasises 'the one condition that would be necessary for such a radical shift to be appropriate: trust in the adjudicators'.[70] As Ginsburg correctly reminds us, particularly the interest-balancing to be carried out under my third principle 'provides little firm anchor for courts in deciding how to proceed',[71] and:

In choosing between a rule and a standard, we need to consider who will be applying the law. Recall that rules (arguably including territoriality as classically formulated)

[67] ibid 75.

[68] Muir Watt states: 'The most challenging question today for international lawyers is the question of the ways in which the very existence of international law (whether public, criminal, administrative or private) in all its conceptual components (jurisdiction, sovereignty, and legal order) is radically threatened by globalization, which transforms processes of law production, displaces authority, neutralises boundaries. Therefore, while a radical conceptual overhaul is certainly long overdue in respect to the extraterritorial assertion of sovereignty, a return to familiar forms of state interest analysis is not necessarily the best way to go about it. I appreciate that Professor Svantesson's proposal is limited to this one point about the desirable re-interpretation of the Harvard Draft principles, and that even while broaching paradigm renewal, he did not intend to wade any deeper into the relationship between international law, public, criminal or private, and global legal theory! However, since he uses the vocabulary of epistemology, it is not inappropriate to respond on the same plane, particularly since this is precisely where debate seems to be heading over the other side of the fence' (ibid 79).

[69] The vague nature of the principles is by no means unique in the context of legal core principles. And, indeed, given that there is a healthy ongoing debate about the true nature of international law (contrast eg the views expressed by Ronald Dworkin ('A New Philosophy for International Law' (2013) 41 Phil & Pub Aff 2) and those of Jack L Goldsmith and Eric A. Posner (*The Limits of International Law* (OUP 2005)), it may perhaps be seen as natural that jurisprudential core principles presented within international law are somewhat vague.

[70] Tom Ginsburg, 'Comment on Svantesson, A New Jurisprudential Framework for Jurisdiction' (2015) 109 AJIL Unbound 86, 88.

[71] ibid 87.

are useful in part because they are simpler to administer. Standards require application of general principles to particular cases, and so require more skill in the adjudicator. Svantesson is proposing a shift from an increasingly unwieldy rule to a looser standard.

As this discussion suggests, a crucial variable in choosing between rules and standards is judicial quality. If we trust the adjudicators to make the right calls, then giving them more discretion in the form of standards is a helpful thing to do. On the other hand, if we think that judges are prone to error, simple rules of thumb will do a better job and lead to fewer mistakes, so long as we think we can formulate a reasonable rule.[72]

All this is correct and an important observation in relation to my proposal. However, I argue that the mechanical simplicity that the proxy principles (at least the central territoriality principle) provide is in actual fact nothing but an illusion. After all, if we cannot agree on *when* an activity takes place in a particular territory—which is the central point of dispute for example in the mentioned matter between Microsoft and the United States government—it is of little comfort that we may agree on the consequences that would follow *if* the activity takes place in a particular territory. And if we lack clear definitions of *where* a state is exercising jurisdiction when it does so (refer to the earlier discussion of the majority opinion in the *Lotus* case), it helps us little that we may agree on the consequences that would follow *if* the state exercises jurisdiction in a particular geographical location.

As observed by Ginsburg: '[W]e are now in a world in which territoriality, which once seemed the master principle of jurisdiction, no longer serves to provide clear and definitive answers in many disputes. The Cyber law cases that motivate Svantesson are only the tip of the iceberg here.'[73] Thus, my claim is this: the move to the core principles will admittedly introduce a broader scope for interpretation, at least until the application of the principles matures, but we are already faced with a comparable degree of vagueness. Therefore, there is no reason to think that the jurisprudential core principles advanced here will cause, for example, a greater risk of overlapping

[72] ibid. Discussing the assessment of legal tests more generally, Ginsburg makes these useful observations: 'Any legal test can be assessed in terms of its ease of application and whether it easily sorts most of the phenomena to be considered into one side or the other. Does it provide relatively clear criteria to resolve most disputes? What is the underlying distribution of cases in the sense that how many of them are hard to resolve? What is the balance of false negatives and false positives that the test is likely to generate? And what are the consequences of those errors? As should be obvious from these questions, a test can not be considered in the abstract, without thinking about who will be applying it' (ibid 86).

[73] ibid.

jurisdiction or negatively impact predictability for the parties, compared to our current, misguided, reliance on territoriality as the jurisprudential core for jurisdiction.

In fact, I would go as far as to suggest that, from a jurisprudential point of view, the framework advocated here represents a welcomed simplification. As noted already, the inadequacy of the territoriality principle is such that it alone cannot constitute the jurisprudential core of jurisdiction; to function, it needs to be backed up by other principles, such as the nationality principle and arguably the other jurisdictional principles outlined in the 'Harvard Draft'. In contrast, under the framework I advance, one will *always* turn to the same three core principles that are applicable in, and relevant for, *every* assessment of jurisdiction.

For example, while the territoriality principle may account for why Germany may claim jurisdiction over my speeding offences there, it cannot accommodate Australia claiming jurisdiction over child sex offences committed by an Australian citizen in another country. In contrast, my three core principles can be used to assess jurisdiction in both of these—from a territoriality perspective different—situations.

In other words, the framework I advance provides for both increased simplicity and for increased predictability, in that we always know in advance which principles will be used to assess jurisdiction, that is, my three core principles. Where adopted, this uniform core of all jurisdictional inquiries ought to ensure that we quite quickly see a body of examples of their application being developed. After all, the application of the three core principles, even where applied through second-level principles built on the core principles, in relation to a situation that under our current paradigm may fall under the territoriality principle, may well help inform their application in another situation that under our current paradigm would fall under, say, the nationality principle. Thus, at least over time, the proposed framework may in fact be easier for judges—also those judges that do not reach Forseti's level of judicial excellence[74]—to work with.

From the discussions above we may perceive the contours of another possible objection to what I am proposing. It could be argued that, even if we accept my core principles, their broad scope will act as an incentive for, if not necessitate, the subsequent development of more specific proxy principles that, while drawing upon the core principles, represent the principles used in practice. This is a valid concern, and I want to make two observations

[74] According to Snorri Sturluson in the *Prose Edda*, Forseti is a Norse god of justice and all disputants at law who bring their cases before him leave perfectly reconciled.

here. First, I do not object to such a development. It is both natural and desirable. However, it is of fundamental importance to note that when such proxy principles develop, they are then based on a clear set of appropriate core principles, and as long as they stay true to those core principles they are merely means for the practical implementation of the core principles. In other words, my proposal should not be seen as an objection to, or obstacle for, the important work carried out by many scholars in trying to articulate specific jurisdictional rules. Rather, all I am suggesting is that the important work they are carrying out ought to be guided by, indeed rooted in, the three core principles advanced here. Second, the 'Harvard Draft' principles also are, as noted, vague and broad in scope, and they have also resulted in the natural development of more precise principles helping in their actual application.[75]

Further, some may object that, like the current international law system described as dysfunctional earlier, the proposed core principles fail to eliminate instances of jurisdictional claims being made by more than one state. However, first of all, the balancing to be exercised in the third step of the proposed framework ought to help minimise instances of such overlap. And second, as is widely recognised, the potential for some overlap and even conflict between jurisdictional claims is a natural part of international law.[76] Here, we may usefully recall Rule 7 derived from the majority opinion in the *Lotus* case: that is, that international law allows concurrent jurisdiction.

When speaking about the proposed framework, I have more than once been met by concern about who is to decide whether a 'substantial connection' and 'legitimate interest' have been established. However, the proposed framework will not alter any decision-making processes or who decides what. All that is done is to change the rules being applied, for example, by the courts in their decision-making processes. Thus, the question of who will decide whether a 'substantial connection' and 'legitimate interest' have been established will be answered in the same manner as we currently answer the question of who decides whether a certain activity takes place on a particular territory.

I have also been met by the somewhat confusing—or perhaps more accurately, confused—claim that a multifactor test for jurisdiction would somehow be too complex and that such a test would be particularly prone to misuse. However, first of all, the law is replete with multifactor tests, and the legal community—and in fact the community more broadly—apply multifactor

[75] Consider the—typically more detailed—rules on jurisdiction found in private international law, such as focus placed on the territory in which a contract was formed.

[76] See eg Mills (n 37) 199.

tests on a daily basis. To see that this is so, we need only recall the multifactor assessments used to determine court competence in the 'targeting' test articulated by the CJEU in *Hotel Alpenhof/Pammer*,[77] or why not the US test for general jurisdiction,[78] (or indeed the *International Shoe* test,[79] for specific jurisdiction)? And maybe the most extreme of them all, we may consider the doctrine of *forum non conveniens*,[80] admittedly not a test for jurisdiction *per se* but for declining jurisdiction, which of course is closely related and indeed a form of multifactor assessment to determine court competence.

Second, it is not exactly the case that territoriality-focused single-factor tests cannot be misused. On one occasion, an Austrian court claimed jurisdiction over a French skier in a child-support dispute, based on the fact that the Frenchman had property in Austria. The property in question was a pair of boxer shorts the skier had, presumably accidentally, left behind in an Austrian hotel room.[81] The claim of jurisdiction on that occasion may well have been completely appropriate. However, to anchor the jurisdictional claim in the territoriality-focused single-factor tests of property in Austria seems to be a clear misuse.

Third, and more generally, the simple reality is that law cannot be drafted in a manner that excludes the risk that it is misused or abused—any legal rule may be misused. And if the choice is between, on the one hand, a territoriality-focused single-factor test that just does not work and, on the other hand, a multifactor test that can more easily be misused, surely the prudent path forward is to embrace the multifactor test and work towards minimising the instances of its misuse (eg through appropriate mechanisms for appeals)?

A more serious attack on the framework I advance could be mounted taking the nature of the Internet as its point of departure. As no one would seriously dispute, the Internet is like a living organism in that it is constantly developing and changing—constant change is here to stay, as the saying goes. In the light of this, the argument could be made that we ought not adjust our laws, rather we ought to change how the Internet operates. To place this suggestion in the context of our discussion, it could be said that instead of abandoning the territoriality principle to fit with the operation of the Internet, we ought to squeeze the operation of the Internet into the

[77] Joined cases C-585/08 and C-144/09.

[78] See Svantesson (n 26) 222–28 and 428–29.

[79] *International Shoe Co v Washington* 326 US 310 (1945). See further Svantesson (n 26) 229–33.

[80] Svantesson (n 26) 484–98.

[81] H Bertram-Nothnagel, 'Enforcement of Foreign Judgments and Arbitral Awards in West Germany' (1976–77) 17 VA J Intl L 393 (footnotes omitted).

framework of the territoriality principle. However, I would like to try and dispute such an idea that I find quite unwholesome.

The main feature of the Internet that causes difficulties for the territoriality principle is that it caters for cross-border interactions. I suspect that few would be willing to sacrifice that fundamental feature on the altar of the territoriality principle. Furthermore, the problems associated with using the territoriality principle as the jurisprudential core of jurisdiction are—as noted earlier—not limited to the Internet context; it just happens to be the case that those problems are particularly obvious in the Internet environment.

Finally, while not a criticism of my proposal as such, it may be pointed out that I have merely stated what our goal should be without giving any indication as to how we gets there—I describe the new paradigm, but I do not comment on how to achieve the paradigm shift. This is no doubt an important and fair observation as the discussion above is largely 'legal-technical', without much consideration of law as a political instrument, or the *realpolitik* that often drives legal change. The reality is, of course, that in the absence of a political will, the paradigm shift is highly unlikely to happen. And it is also a reality that law, not least international law, typically moves at a glacial speed. Consequently, there is no reason to think that the paradigm shift will happen overnight even if it should happen to be accepted in the legal community.

The aim of my writings on this is humble—all I am aiming to do is to raise the issue and to advocate a particular solution. Hopefully, this can help put the matter on the agenda and create a legal, and political, discussion that over time can facilitate the sought-after paradigm shift. Thus, I suspect it will be some time before the writers of introductory texts on public international law need to revise their chapters on jurisdiction. But then again, to a great extent, law in general, and international law in particular, is in many ways nothing more than a commonly shared idea, a *folie à deux*, a shared psychosis—international law is what we think it is. And then the first step to change is that we alter how we view the issue of jurisdiction in international law.

Thus, it is for us all—industry, government, courts, international organisations, civil society, and the academic community—to help achieve useful change. Organisations working on, or with, international instruments in the field ought to consider how they can incorporate the proposed framework—directly or by way of using it as guidance—to alleviate the complications caused by a misguided adherence to strict territoriality. Courts and other lawmakers must stop uncritically letting the territoriality principle guide their approaches. And the same applies to the academic community—the body tasked, to a greater degree than any other, with thinking critically. Furthermore, those engaged in capacity-building initiatives must recognise that they need to incorporate capacity building for a sound understanding of the jurisdictional challenges and solutions.

3.5 Merging Public and Private International Law

In his interesting book *Making Laws for Cyberspace*, Reed observes:

> Drafting a law so as to limit its online application is not a simple task. It is necessary to produce a definition of those persons to whom it should apply which, at the same time, excludes online actors who are not intended to fall within the scope of the law. The solution seems to lie in identifying an appropriate link to the territory of the lawmaker. Over the years, private international law has developed for this purpose.[82]

The framework I advanced earlier is indeed aimed at ensuring that laws apply where they should and do not apply where they should not. However, as is already clear, I cannot agree with Reed that the solution is to be found in territoriality. Neither do I agree with the characterisation of private international law rules as having developed for the purpose of identifying appropriate links *to the territory* of the lawmaker. I think Reed could have said what he says here without linking his claim to territoriality. Indeed, I wish here to emphasise my suspicion that the territoriality-nihilistic jurisprudential framework for jurisdiction presented, with its three core principles, may also be viewed as a useful representation of the core principles of private international law (or conflict of laws as it is also called). In other words, I argue that the core of private international law is also to be found in principles other than territoriality.

Granted this, the principles I propose could be seen not only to hint at a convergence between public and private international law, but in fact also to constitute the common core that indeed unites public international law and private international law. Uniting the two areas of law into one discipline—a discipline covering both what traditionally falls within private international law, and the jurisdictional issues that arise within public international law—seems like a both natural and desirable development.[83] After all, the jurisdictional issues that arise within public international law have a lot more in common with the jurisdictional issues that arise within

[82] Chris Reed, *Making Laws for Cyberspace* (OUP 2012) 39 (internal footnote omitted).

[83] As noted by Ryngaert (n 9) 84, 'The ensuing entwining of public and private international law is not new. It heralds a return to the early 19th century, when, as the work of Joseph Story testifies to, both fields were not yet strictly separated.' ('An Urgent Suggestion to Pour Old Wine Into New Bottles: Comment on "A New Jurisprudential Framework for Jurisdiction" '.)

private international law than they do with many other matters falling within public international law, such as the principles for how we should interpret international treaties. As stated by McDougal and Jasper, both public and private international law are aimed at 'allocating among states of the world the competence to make and apply law to the transnational events that affect them'.[84] Indeed, as pointed out by Nagan in a particularly insightful article from the early 1980s:

> From a functional point of view, the distinction between public and private international law would appear to be at best artificial, as both public and private international law ultimately deal with the myth and practice of responding to claims for the allocation of the good as well as the undesirable things in the world social processes. [...] [P]ublic and private international law are in reality complementary and indispensable components of a larger and more inclusive conception of world public order.[85]

I have no doubt that the jurisdictional challenges that face the Internet, as well as the jurisdictional challenges facing our modern world more generally, would be much better, and more efficiently, addressed if we managed to break down the illusion of a dividing barrier between jurisdictional issues arising with private international law and those arising within public international law.[86] The future will tell, and just how far apart private and public international lawyers currently are is well illustrated by the perhaps trivial fact that both disciplines use the abbreviation 'PIL' to refer to their respective disciplines, seemingly unaware—or at least with total disregard for the fact—that the other discipline uses it as well![87]

At any rate, as I have pointed to elsewhere,[88] the factors taken into account in private international law—both in Civil Law and Common Law—can be broken down into eleven categories, as follows.

[84] M McDougal and R Jasper, 'The Foreign Sovereign Immunities Act of 1976: Some Suggested Amendments' in M Landwehr (ed), *Private Investors Abroad—Problems and Solutions in International Business in 1981* (1982) 6, as referred to in WP Nagan, 'Conflicts Theory in Conflict: A Systematic Appraisal of Traditional and Contemporary Theories' (1981–82) 3 J Intl & Comp L 343, 367.

[85] Nagan (n 84) 361.

[86] As correctly pointed out by Muir Watt (n 8) 75), 'This rejection [of the public/private distinction in international law] has been championed both as a normative matter, because so much would be gained from reaching 'beyond the schism' to overcome the personality split from which the discipline has suffered for over a century, as well as a descriptive matter, since multiple issues, debates, myths and concepts straddle the divide, even if they give rise to specific disciplinary understandings on either side.'

[87] If we are to continue treating these two disciplines as separate, perhaps we could agree to abbreviate private international law as 'PrIL' and use the abbreviation 'PuIL' for public international law.

[88] Svantesson (n 26) 424.

Factors relating to:

1. the interests of the forum jurisdiction;
2. the interests of other jurisdictions;
3. the interests of the plaintiff;
4. the interests of the defendant;
5. the interests of any relevant third-parties;
6. the plaintiff's contacts with the forum jurisdiction;
7. the defendant's contacts with the forum jurisdiction;
8. the plaintiff's contacts with other jurisdictions;
9. the defendant's contacts with other jurisdictions;
10. the dispute's contacts with the forum jurisdiction; and
11. the dispute's contacts with other jurisdictions.

These factors from the different categories can be combined, balanced, and weighted in an infinite number of ways. For example, equally well as stating that the courts shall claim jurisdiction over tort disputes if the tort was committed in the forum state (ie focusing on an aspect of the dispute's contacts with the forum), one could decide that the courts will hear such disputes only where the plaintiff is domiciled in the forum state (ie focusing on an aspect of the plaintiff's contacts with the forum). This highlights the great potential for a diversity of approaches.

In our context here, the important question is whether all of these factors can be fitted within the three core principles I proposed. If they cannot, then clearly the three core principles, whatever their value within public international law, cannot constitute the core of the current private international law thinking.

Significantly, while, as illustrated by the above list, some private international law rules are rather closely aligned with territoriality,[89] I have not been able to think of any jurisdictional rules in private international law that do not rely on one or more of the core principles I have articulated. I argue that they do indeed fit, and fit well, as is illustrated in Table 3.1.

To this we may add the fact that other private international law 'tools' also fit well within the proposed jurisprudential framework. Consider, for example, the Common Law doctrine of *forum non conveniens*. It clearly invites the type of interest balancing that is emphasised in the jurisprudential framework I advance here.

[89] Crawford (n 6) 474.

Table 3.1 Factors in private international law

Substantial connection Factors relating to:	**Legitimate interest Factors relating to:**	**Balance of interests Factors relating to:**
the plaintiff's contacts with the forum jurisdiction	the interests of the forum jurisdiction	the interests of other jurisdictions
the defendant's contacts with the forum jurisdiction		the interests of the plaintiff
the dispute's contacts with the forum jurisdiction		the interests of the defendant
		the interests of any relevant third parties
		plaintiff's contacts with other jurisdictions
		the defendant's contacts with other jurisdictions
		the dispute's contacts with other jurisdictions

These observations lend support to the proposition that what I put forward as the three core principles are in fact such core principles in both public and private international law. At a minimum, it demonstrates that the core principles I advance can indeed accommodate the categories of factors relevant in private international law, in a way that allows us to analyse the current private international law rules based on the proposed jurisprudential framework.

Such an analysis will no doubt highlight weaknesses in some of our current private international law rules. To give but one example, we can consider how the previously mentioned jurisdictional rule that allows jurisdiction to be based on the fact that a contract was formed in the territory of the state exercising jurisdiction[90] would fare if analysed based on the proposed jurisprudential framework. Imagine, for example, that a contract is concluded online between two Danes, both of whom are domiciled and habitually residing in Denmark. Imagine further that, at the time of contracting, one of the parties to the contract happened to be travelling through Australia, and that after several rounds of emails exchanging

[90] For England, see Civil Procedure Rules 1998 Practice Direction 6B para 3.1(6)(a). For Australia, see M Davies, A Bell, and P Brereton, *Nygh's Conflict of Laws in Australia* (9th edn, LexisNexis Australia 2014) 55–57. Valid in all states and territories. For Hong Kong, see The Rules of the High Court (Ch 4A)—Order 11(1)(d)(i). For China, see Civil Procedure Law of the People's Republic of China, Art 265.

counter offers, the contract is finally deemed to have been formed in Australia. Under Australian law, an Australian court could now claim jurisdiction should a dispute arise.

While there is some form of connection to Australia as the contract was 'formed' there, that connection can hardly be described as substantial. Furthermore, it is difficult to see how Australia can claim some form of legitimate interest in the dispute.[91] Finally, it seems clear that there are several other interests that, on balance, would weigh heavier in the analysis under my third principle. Indeed, I would argue that, in analysing a situation such as this we would not even reach the third principle as there is no legitimate Australian interest.

3.6 Concluding Remarks Regarding the Proposed Framework

Abandoning the territoriality principle and the other 'Harvard Draft' principles represents a paradigm shift and will no doubt be associated with controversy and opposition. Given how long we have been following the grooves of habitual reliance on the territoriality principle, my proposal will no doubt need to spend time in fermentation before it is palatable to a broader audience. However, in the debates to be had about this, it is of fundamental importance to recall the limits the drafters themselves had in mind for the 'Harvard Draft' principles; they were articulated in the penal context representing a simplified restatement of state practice and academic opinions at the time. Further, account may be taken of the fact that the 'Harvard Draft' principles are arguably merely proxies for underlying core principles. Thus, returning to the three core principles advanced here may perhaps be surprisingly natural, not least as the territoriality principle and the other 'Harvard Draft' principles can remain operational as second-level principles. Thus, to put it in gardening terms, while my proposal represents a jurisprudential land clearance and replanting, on a practical level it is more of a careful, but important, pruning exercise than it is deforestation.

In addition, it may be noted that, while the use of territoriality as the jurisprudential core principle for jurisdiction demonstrably has not stood the test

[91] I acknowledge that if the contract is illegal under Australian law and relates to some particularly offensive conduct, Australia could arguably be seen to have an interest in prosecuting the parties to the contract. However, that is a separate matter to the jurisdiction over the civil matter discussed here.

of time, I suspect that the three principles I advance are as 'future proof'[92] as legal principles can be and that they are likely to 'reflect relative stability in expectation about the allocation of power or competence to make and apply law to events, persons, and values over which control is shared by more than one state'.[93]

If we accept the three principles I advance as the common core of jurisdiction both under public international law and private international law, we have also managed to bridge the gap too often perceived to separate these two areas of law and bring them together into one discipline. Such a step opens up the doors for a more productive study and discussion of these issues of fundamental importance both in public and in private international law.

I have emphasised that, as the name makes clear, the 'core principles' I propose are merely the core from which layers of more specific and detailed legal principles and rules should be constructed. Given the fact that we already have established systems of rules for jurisdiction, one important task that lies ahead is to examine those existing rules from the perspective of the three core principles. What we need is nothing less than a serious attempt at a taxonomic classification of all our jurisdictional norms and principles akin to that undertaken by Swedish scientist Carl von Linné (*Carolus Linnaeus*) in relation to plants in the 1700s.[94] This will no doubt be a massive task, and it not a task I wish to embark upon here.

Finally, while I have already quoted Mills extensively, I want to include one last quote from his work highlighting just how significant the proposed paradigm shift is:

> Law and the social sciences are fundamentally different from the natural sciences in that the adoption of a theoretical perspective does not merely describe, but may also change its subject. The move from Newtonian to relativistic physics did not change the reality of the world, it simply described it better. But the development of classical international law did not merely describe movements in international relations, it has helped to shape them by shaping the thinking and behaviour of the actors who in turn influenced events. The choice of a theoretical paradigm in law is

[92] A highly useful concept with a history in other scientific fields and in marketing (see eg <http://www.uhasselt.be/documents/tijdschriften/nuweetjehet/nuweetjehet-04-0910.pdf>, accessed 3 April 2017), to my knowledge first used by Chris Reed in the context of legal principles.

[93] Nagan (n 84) 343, 372.

[94] Carl von Linné, *Systema naturæ per regna tria naturæ, secundum classes, ordines, genera, species, cum characteribus, differentiis, synonymis, locis* (10th edn, 1758).

not only a question of its descriptive accuracy, but also a question of its normative implications.[95]

Thus, the advance in theory that the proposed territoriality-nihilistic jurisprudential framework for jurisdiction provides for some very interesting future developments in relation to the crucially important concept of jurisdiction.

[95] Mills (n 37) 237.

4

A Very Brief History of Internet Jurisdiction

Any account of history ought to start with the caveat implied in Napoleon Bonaparte's observation that 'History is the version of past events that people have decided to agree upon.'[1] Nevertheless, elsewhere, I have argued that the history of Internet jurisdiction so far may be divided into four relatively distinct phases, and that these phases oscillate between over-regulation and under-regulation. In other words, it is my view that the regulators' approaches towards the issues that broadly fall within the field of Internet jurisdiction move in a pendular fashion between the mentioned extremes. Here I wish to approach this issue rather from the perspective of providing a brief, and no doubt too simplistic, account of the history of Internet jurisdiction by highlighting certain key events. In doing so, I hope to prove one point and emphasise another. I hope to prove my claim of four relatively distinct phases of attitudes since Internet jurisdiction first gained attention, and I hope to emphasise the point I also have made that no razor-sharp lines can be drawn between these phases.

Finally, I use this chapter to start lining up against the wall some of my favourite 'mythical solutions': my pet hates, in the Internet jurisdiction area. Throughout the book, I aim to then shower them in what I hope are sharp destructive arguments once and for all relegating them to their well-deserved place in the rubbish bin.

4.1 In the Beginning There Was Nothing (pre-1991)

The Internet predates Internet jurisdiction issues, and jurisdictional rules predate the Internet. These are of course rather rudimentary observations, but ones that still ought to be noted. In the earliest days of Internet, and the pre-Internet networks that led to the Internet, the discussions were not focused on the jurisdictional issues that now concern us so greatly. The only

[1] <https://www.brainyquote.com/quotes/quotes/n/napoleonbo101520.html>, accessed 3 April 2017.

Solving the Internet Jurisdiction Puzzle. First Edition. Dan Jerker B. Svantesson.

reason this deserves to be mentioned is that we must always remind ourselves that the Internet was not designed with jurisdictional conundrums in mind. Therefore, we must never fool ourselves into assuming that any of the Internet's core structures are intended to affect Internet jurisdiction questions in one direction or another.

Of equal, or indeed greater, significance, to understand the jurisdictional rules that currently apply online, we need to acknowledge their origin in the jurisdictional thinking that predates the Internet; because the modern high-tech Cyberworld is largely governed by very old jurisdictional rules—there is a clash between the modern Cyber-landscape and the jurisdictional dinosaur rules sought to be applied to it.

4.2 The Wild Wild Web (1991–99)

In 1991, CERN researcher Tim Berners-Lee developed the World Wide Web (WWW). While clearly there was an Internet prior to 1991, and while that pre-1991 Internet did give rise to legal issues (including of a jurisdictional kind), I will here use 1991 as the starting point of the first phase—a phase—where the Internet was approached somewhat like *terra nullius*—that stretched to the millennium shift. During this era, there were significant calls for states to refrain from making jurisdictional claims over the Internet and Internet-related activities, and the US government was leading the way with 'self-regulation' for the so-called 'information superhighway'.

Particularly the early years of this period were characterised by great hope; with people dreaming of the WWW-driven Internet as a new frontier, open to everyone and regulated by no one. It was initially an era of freedom during which the abbreviation WWW might just as well stood for the 'Wild Wild Web'. A defining moment took place in 1996 when John Perry Barlow famously unveiled his ground-breaking 'Declaration of the Independence of Cyberspace': 'Governments of the Industrial World, you weary giants of flesh and steel, I come from Cyberspace, the new home of Mind. On behalf of the future, I ask you of the past to leave us alone. You are not welcome among us. You have no sovereignty where we gather …'[2] The somewhat enigmatic quality of this 'Declaration' may make it look amusingly eccentric, or even utterly absurd, today. However, it must be remembered that it was presented to a world very different from the world of today, at least far as Internet use is

[2] JP Barlow, *A Declaration of the Independence of Cyberspace* (8 February 1996) <https://projects.eff.org/~barlow/Declaration-Final.html>, accessed 3 April 2017.

concerned. We have since long passed the point at which it was still possible to differentiate between the 'real' bricks and mortar world, on the one hand, and the 'virtual' world of Cyberspace on the other. Twenty years ago, people would say things like 'this afternoon I am going online for an hour'. Today, due to our smart-metered homes, our constant presence on social media, and not least due to 'man's best friend'—our smart phones—the distinction between Cyberspace and the 'real' world is very much blurred. And attempting to draw such a distinction will not be possible for legal purposes. Thus, while it may be a 'silver bullet' in relation to the Internet jurisdiction puzzle, making the Internet be beyond the law is clearly an impossible solution.

In dealing with Internet jurisdiction cases, courts were showing great creativity in many cases. Looking back at the decisions from this time—especially amongst some important US courts—one almost gets the feeling of there being a 'race' to be the first to find the solution to the Internet jurisdiction conundrum.[3] Unfortunately, comparatively speaking, there is precious little of this entrepreneurial spirit left in the court decisions of today.

The most influential decision of this first phase is no doubt the 1997 case of *Zippo Manufacturing Company v Zippo Dot Com, Inc.*[4] In fact, one commentator describes it as 'the first compelling analysis of personal jurisdiction related to the Internet',[5] while another states that this is the 'dominant test for determining whether a web site is subject to personal jurisdiction'.[6] While it no longer maintains its dominance, the *Zippo* case remains of importance.

In the *Zippo* case, the Court drew a distinction between different types of website and divided Internet presence into three separate categories:

[3] For more details on some of the seminal cases, refer to the excellent study by the Fordham Center on Law and Information Policy noting: 'Specifically, these jurisdictional questions appear most frequently within the Ninth Circuit, which heard more than twice the number of Internet jurisdiction cases than any other federal circuit. Approximately 25 percent of the 41 seminal cases identified originate from the Ninth Circuit. The second most common circuit for Internet jurisdiction cases is the Second Circuit. Conversely, there are comparably few Internet jurisdiction cases from the First Circuit, the District of Columbia Circuit or the Federal Circuit, suggesting that these are less important circuits for Internet jurisdiction matters. With respect to the remaining federal circuits, there appears to be a relatively equal distribution of Internet jurisdiction cases among them' (Joel R Reidenberg et al, 'Internet Jurisdiction: A Survey of Legal Scholarship Published in English and United States Case Law' (30 June 2013) Fordham Law Legal Studies Research Paper No 2309526 <http://ssrn.com/abstract=2309526 or http://dx.doi.org/10.2139/ssrn.2309526> 55, both accessed June 2017).

[4] 952 FSupp 1119 (WD Pa 1997) ('*Zippo*').

[5] JM Jensen, 'Personal Jurisdiction in Federal Courts over International E-commerce Cases' (2006–07) 40 Loy LAL Rev 1507, 1514.

[6] KF King, 'Personal Jurisdiction, Internet Commerce, and Privacy: The Pervasive Legal Consequences of Modern Geolocation Technologies' (2011) 21 Albany L J Sci & Tech 61, 63.

> [O]ur review of the available cases and materials reveals that the likelihood that personal jurisdiction can be constitutionally exercised is directly proportionate to the nature and quality of commercial activity that an entity conducts over the Internet. This sliding scale is consistent with well developed personal jurisdiction principles. At one end of the spectrum are situations where a defendant clearly does business over the Internet. If the defendant enters into contracts with residents of a foreign jurisdiction that involve the knowing and repeated transmission of computer files over the Internet, personal jurisdiction is proper. At the opposite end are situations where a defendant has simply posted information on an Internet Web site which is accessible to users in foreign jurisdictions. A passive Web site that does little more than make information available to those who are interested in it is not grounds for the exercise of personal jurisdiction. The middle ground is occupied by interactive Web sites where a user can exchange information with the host computer. In these cases, the exercise of jurisdiction is determined by examining the level of interactivity and commercial nature of the exchange of information that occurs on the Web site.[7]

The 'sliding scale' of interactivity of the *Zippo* case has played a central role in US caselaw for a number of years.[8] However, while its influence can still be seen today,[9] it is important to point out that the *Zippo* test has also been criticised by some courts,[10] and even where it is applied, its application is not uniform.[11]

The lasting jurisprudential value of the *Zippo*'s sliding scale is the very fact that it points to a scale: it expressly recognises that what we are dealing

[7] *Zippo Manufacturing Company v Zippo Dot Com, Inc* 952 FSupp 1119 (WD Pa 1997).

[8] JM Jensen, 'Personal Jurisdiction in Federal Courts over International E-commerce Cases' (2006–07) 40 Loy LAL Rev 1507, 1515. See also *Carlson v Fidelity Motor Group, LLC* 360 Wis 2d 369 (2015), where the court declined to apply *Zippo*, finding that the *Hy Cite Corp* case 'provides a better approach by not looking to a separate test where Internet websites are involved, but rather simply considering such websites as a part of the overall due process question of "whether the defendant's contacts with the state are of such a quality and nature such that it could reasonably expect to be haled into the courts of the forum state"' (at 304, FN4, quoting *Hy Cite*, 297 F Supp 2d, 1161). This court said it was persuaded by the observation of another federal court that declined to follow the *Zippo* analysis: '[T]his court observes that the need for a special Internet-focused test for "minimum contacts" has yet to be established. It seems to this court that the ultimate question can still as readily be answered by determining whether the defendant did, or did not, have sufficient "minimum contacts" in the forum state. The manner of establishing or maintaining those contacts, and the technological mechanisms used in so doing, are mere accessories to the central inquiry […]' citing *Winfield Collection, Ltd v McCauley*, 105 F Supp 2d 746, 750 (ED Mich 2000).

[9] Consider eg *Best Odds Corp v iBus Media Ltd*, No 13-2008, slip op (D Nev 15 May 2014) 10.

[10] See eg *Oldenfield v Pueblo De Bahia Lora, SA*, 558 F3d 1210, 1219 (11th Cir 2009) and *Best Van Lines, Inc v Walker*, 490 F3d 239, 252 (2d Cir 2007).

[11] See further KF King, 'Personal Jurisdiction, Internet Commerce, and Privacy: The Pervasive Legal Consequences of Modern Geolocation Technologies' (2011) 21 Albany LJ Sci & Tech 61, 82–86.

with—the legitimacy of jurisdictional claims over online activities—is a matter of degree rather than a binary matter; an important insight that seems then to have become lost along the way since the *Zippo* case was decided some twenty years ago.

And speaking of lost insights, it seems to me that courts today have something to learn from the self-awareness displayed by the Court in another decision from 1997; that is, *American Libraries Association v Pataki*: 'Judges and legislators faced with adapting existing legal standards to the novel environment of cyberspace struggle with terms and concepts that the average [. . .] five-year-old tosses about with breezy familiarity.'[12]

Looking at the academic discourse of this period, several things are noteworthy. Like many of the court decisions from the same period the most striking characteristic of the academic discourse is its openness to debating even the fundamentals. There was creativity and a willingness, indeed eagerness, to find innovative solutions.

Consider, for example, Menthe's innovative—at the time—idea that the Internet is a separate space akin to other international spaces such as the High Seas, Antarctica, and Outer Space (ie as an independent international space beyond individual nations' regulation).[13] Or why not Reidenberg's important work on *Lex Informatica*,[14] and of course, Lawrence Lessig's well-known work.[15]

However, looking at jurisdictional issues specifically, the most significant literature was doubtless that created by the most interesting academic 'exchanges of ideas' to date in the field of Internet jurisdiction. I am here referring to the series of articles written by Johnson and Post, on the one hand (taking a Cyber-libertarian point of view), and Jack Goldsmith's writings on the other hand.

In 1996, Johnson and Post wrote an article titled 'Law And Borders—The Rise of Law in Cyberspace'[16] in which they argued that the Internet should be viewed as a separate 'space',[17] beyond the control of individual nations' regulation. Moreover, the article suggested that, to the extent that this separate space is to be regulated, such regulations would emerge in the form of

[12] [1997] SDNY 969 F Supp 160, 170 (per Preska J).

[13] DC Menthe, 'Jurisdiction in Cyberspace: A Theory of International Spaces' (1998) 4 Mich Telecom & Tech L Rev 69, 85.

[14] J Reidenberg, 'Lex Informatica' (1998) 76(3) Tx L Rev 553.

[15] L Lessig, 'The Law of the Horse: What Cyberlaw Might Teach' (1999) 113 Harv L Rev 501; L Lessig, *Code and Other Laws of Cyberspace* (Basic Books 1999).

[16] David Johnson and David Post, 'Law and Borders—The Rise of Law in Cyberspace' (1996) 48 Stan L Rev 1367.

[17] ibid 1378.

self-regulation.[18] Johnson and Post's reasoning is neatly summarised in this statement:

> Treating Cyberspace as a separate 'space' to which distinct laws apply should come naturally, because entry into this world of stored online communications occurs through a screen and (usually) a 'password' boundary. There is a 'placeness' to Cyberspace because the messages accessed there are persistent and accessible to many people. You know when you are 'there'. No one accidentally strays across the border into Cyberspace. To be sure, Cyberspace is not a homogenous place; groups and activities found at various online locations possess their own unique characteristics and distinctions, and each area will likely develop its own set of distinct rules. But the line that separates online transactions from our dealings in the real world is just as distinct as the physical boundaries between our territorial governments—perhaps more so.
>
> Crossing into Cyberspace is a meaningful act that would make application of a distinct 'law of Cyberspace' fair to those who pass over the electronic boundary.[19]

Essentially replying to this article, Goldsmith published 'Against Cyberanarchy' in 1998. In that article, he emphasised that:

> Cyberspace transactions are no different from 'real-space' transnational transactions. They involve people in real space in one jurisdiction communicating with people in real space in other jurisdictions in a way that often does good but sometimes causes harm. There is no general normative argument that supports the immunization of cyberspace activities from territorial regulation. And there is every reason to believe that nations can exercise territorial authority to achieve significant regulatory control over cyberspace transactions.[20]

A few additional shots, such as Post's interesting reply 'Against "Against Cyberanachy"',[21] were fired in this exchange; however, none as significant as the opening salvos mentioned.

4.3 The First Phase of Over-regulation (2000–09)

Around the time the world was preoccupied by the horrible events that computers around the world were meant to cause in response to the new millennium, signs emerged of a significant change in attitudes towards Internet jurisdiction.

[18] ibid 1367. [19] ibid 1379 (footnotes omitted).

[20] JL Goldsmith, 'Against Cyberanarchy' (1998) 65 U Chicago L Rev 1250.

[21] David Post, 'Against "Against Cyberanarchy"' (2002) 17 Berkeley Tech LJ 1365.

By the start of the new millennium, courts and legislators had developed a rather aggressive attitude as to when they could claim jurisdiction over Internet conduct. Put simply, their attitude was that their jurisdictional powers extended to any Internet conduct that impacted, or had the potential to impact, on their territory or citizens. The global nature of the Internet, combined with a limited utilisation of geographical identifiers, meant that courts could claim jurisdiction over virtually all Internet conduct. Or, looked at from the perspective of someone who makes content available online, they could come under the jurisdiction of any court in the world and should, thus, abide by all the laws of all the countries in the world. The impracticalities of this are now well understood and need not be elaborated upon.

There is no lack of cases that illustrate the complications that this situation gives rise to. I will use two of the most prominent cases as examples. The first is the December 2002 Internet defamation decision by the High Court of Australia.[22] The Court allowed Victorian businessman, Joseph Gutnick, to sue US publishing company Dow Jones & Company Inc, in a Victorian court over an allegedly defamatory article available on Dow Jones's website.[23] Further it was decided that Victorian law would be applied. Characteristically for the period, the majority of the Court stated:

> However broad may be the reach of any particular means of communication, those who make information accessible by a particular method do so knowing of the reach that their information may have. In particular, those who post information on the World Wide Web do so knowing that the information they make available is available to all and sundry without any geographic restriction.[24]

In other words, you enter the online environment at your own risk as far as Internet jurisdiction is concerned. One of the flawed assumptions underpinning the approach articulated in the quote immediately above is that it equates an online presence with an (objective) intention to reach the world at large. To see that this is not so, one needs just consider whether a small pizza restaurant in Tyringe (Sweden) really can be said to express an (objective) intention to reach the world at large merely by making its menu available on

[22] *Dow Jones & Company Inc v Gutnick* (2002) 210 CLR 575. This case has been strongly influential both in Australian courts and abroad. Furthermore, it has been subjected to extensive and ongoing academic scrutiny. See eg D Rolph, 'Publication, Innocent Dissemination and the Internet after Dow Jones & Co Inc v Gutnick' (2010) 33 UNSWLJ 562.

[23] *Dow Jones v Gutnick* [2002] 210 CLR 575.

[24] *Dow Jones v Gutnick*, 605.

a website. Suggesting that this is so would be simply ridiculous. The Court has here lost sight of the use of the WWW to interact with a local market.

As observed by Slane, it is important to bear in mind that, in adopting this approach, the Court 'did not assume jurisdiction merely because the defamatory article was available on websites, but rather because it was actually accessed in Australia by Australian subscribers'.[25] This difference between, on the one hand, mere theoretical accessibility and, on the other, actual access, is of course of significance. At the same time the distinction is not always as clear as a superficial examination may suggest. After all, in many Internet defamation cases, it is highly problematic to prove the actual number of times the defamatory content has been accessed. And where such evidence is lacking, the practical distinction between actual access and mere accessibility is not quite so clear.

The *Gutnick* case is also illustrative of a judicial hesitation to treat the Internet as something novel. In a now classic statement, Callinan J took the view that '[t]he Internet, which is no more than a means of communication by a set of interconnected computers, was described, not very convincingly, as a communications system entirely different from pre-existing technology.'[26]

This era—the first phase of over-regulation—also provided us with the most well-known Internet jurisdiction matter of all time (well, at least, so far). The transatlantic dispute between US Internet company Yahoo! and two French associations, La Ligue Contre le Racisme et L'Antisemitisme and L'Union des Étudiants Juifs de France, related to Yahoo!'s operation of a website which, amongst other things, contained an auction service where Nazi memorabilia/junk was frequently on offer.[27] The website could be described as the Yahoo! family's 'flagship', and in contrast to the country-specific Yahoo! sites (eg www.yahoo.fr), this site was said to be aimed at the world at large.[28] When La Ligue Contre le Racisme et L'Antisemitisme et al attempted to have Yahoo! remove the Nazi material from the auction service, in accordance with French penal Code,[29] Yahoo! refused.

[25] A Slane, 'Tales, Techs, and Territories: Private International Law, Globalization, and the Legal Construction of Borderlessness on the Internet' (2008) 71 L & Contemp Prob 129, 142.

[26] *Dow Jones & Company Inc v Gutnick* [2002] HCA 56 [180]. In sharp contrast, see Kirby J's statement in the same case: 'Intuition suggests that the remarkable features of the Internet (which is still changing and expanding) makes it more than simply another medium of human communication. It is indeed a revolutionary leap in the distribution of information, including about the reputation of individuals' (*Dow Jones & Company Inc v Gutnick* [164]).

[27] However, the auction service was not at all specifically designed for the purpose of auctioning Nazi material.

[28] A notion backed by the fact that country-specific advertisement was provided on the site.

[29] *French Penal Code 1791* Art R645-1.

The French Court characterised Yahoo!'s activities as a tort (*faute*) and issued a civil law injunction based on the French Code of Civil Procedure. The Court ruled that Yahoo! must take steps to prevent French Internet users from accessing the sections of the auction site containing Nazi memorabilia.[30]

These cases, like many others, share an important characteristic—they highlight a gap between what can be regarded as legitimate grounds for claiming jurisdiction and what can be regarded as legitimate grounds for expecting another country to recognise and enforce the judgment. In other words, in both these cases it could be argued to be legitimate for the courts in question to claim jurisdiction over the dispute. At the same time, it can also be argued that both cases involve scenarios in which it is perfectly reasonable to refuse to recognise and enforce the judgments.

In the *Yahoo* case, that is exactly what happened. Yahoo! sought and obtained a summary judgment to the effect that US courts would not enforce the French decision. While acknowledging France's right to make law for France, J Fogel decided in Yahoo!'s favour, granting the summary judgment, declaring that the 'First Amendment precludes enforcement within the United States of a French order intended to regulate the content of its speech over the Internet'.[31]

The *Gutnick* case was settled and there was consequently no need for Mr Gutnick to seek to have a judgment enforced in the US. Had he had reason to do so, one can expect that he would have run into difficulties, as US courts have a tradition of refusing to recognise and enforce foreign judgments that are viewed as out of line with strict approach to freedom of speech of the US.

Turning to the academic commentary of this period, it is hard to go past Geist's insightful—and cleverly titled—article, 'Is There a There There? Towards Greater Certainty for Internet Jurisdiction',[32] which brings attention to the inadequacies of the *Zippo* passive versus active test and instead promotes the move towards a targeting-based analysis. This period also saw the publication of the first major monographs specifically addressing Internet jurisdiction,[33] as well as Goldsmith and

[30] *International League Against Racism & Anti-Semitism (LICRA) v Yahoo! Inc* [2000] County Court of Paris.

[31] *Yahoo! Inc v La Ligue Contre le Racisme et L'Antisemitisme* [2001] NDCal 169 F Supp 2d 1181, 22. See, also, *Yahoo! Inc v La Ligue Contre le Racisme et L'Antisemitisme* [2006] 9th Cir 433 F3d 1199.

[32] Michael Geist, 'Is There a There There? Towards Greater Certainty for Internet Jurisdiction' (2001) 16 Berkeley Tech LJ 1345.

[33] ZS Tang, *Electronic Consumer Contracts in the Conflict of Laws* (Hart 2009); U Kohl, *Jurisdiction and the Internet: Regulatory Competence over Online Activity* (CUP 2007); H Spang-Hanssen, *Cyberspace & International Law on Jurisdiction* (DJØF 2004); LE Gillies, *Electronic*

Wu's influential book *Who Controls the Internet? Illusions of a Borderless World*.[34]

4.4 A Degree of Under-regulation (2010–14)

From around 2010, we see a return of under-regulation. And during this third phase, courts and legislators around the world seemed to have started to accept, to a degree, the impossibility of viewing an online presence as an indication of an intention to do business with the world at large. This is of course helpful *per se*. However, it seems there was a tendency to be so over-eager to avoid too broad jurisdictional claims as to end up with a degree of under-regulation. A decision of the European Court of Justice (ECJ) is particularly illustrative.

In 2010, the ECJ addressed the two cases of *Pammer v Reederei Karl Schlüter GmbH & KG*[35] and *Hotel Alpenhof GesmbH v Oliver Heller*[36] jointly. The key question common to the two cases was whether the fact that a website can be consulted on the Internet in the Member State of the consumer's domicile is sufficient to justify a finding that commercial or professional activities are being directed to that Member State within the meaning of Article 15(1)(c) of Regulation No 44/2001. The Court concluded:

- The mere fact that a website can be accessed in the consumer's jurisdiction does not mean that the business has directed its activities to that state;[37] and
- Whether a trader has directed its activity to the Member State of the consumer's domicile, should be ascertained by reference to 'whether, before the conclusion of any contract with the consumer, it is apparent from those websites and the trader's overall activity that the trader was envisaging doing business with consumers domiciled in one or more Member States, including the Member State of that consumer's domicile, in the sense that it was minded to conclude a contract with them.'[38]

The first of these conclusions, while sensible as such, is irrelevant for the cases at hand. Both the *Pammer* case and the *Hotel Alpenhof* case involved

Commerce and International Private Law: A Study of Electronic Consumer Contracts (Ashgate 2008); Dan Svantesson, *Private International Law and the Internet* (Kluwer Law International 2007).

[34] Jack Goldsmith and Tim Wu, *Who Controls the Internet? Illusions of a Borderless World* (OUP 2008).

[35] [2008] Case C-585/08.

[36] [2009] Case C-144/09.

[37] [2010] joined Cases C-585/08 and C-144/09, para 95.

[38] ibid.

contractual situations, as can be expected when the disputes relate to the proper interpretation of an Article addressing contracts concluded by consumers. Thus, we are not dealing here with the more or less randomness of contact and risk typical of claims relating for example to marketing practices or other non-contractual situations.

In other words, it must be remembered that the only situation in which the matter of whether a business has directed its activities to the consumer's state arises is where that business has entered into a contract with the consumer. To me, it seems contrary to intuition to conclude that a business that has contracted with a consumer did not, at least *prima facie*, direct its activities to the consumer's state.[39] Thus, the existence of a contract between the business and the consumer should give rise to a rebuttable presumption that the business directed its activities to the consumer's state. That is, in every case where a business has contracted with a consumer in another Member State, it should be presumed that the business has directed its activities to that Member State. After all, the business has made the choice to reap the benefits of contracting with a consumer from that Member State. My suggestion could consequently be described as an approach focused on 'dis-targeting', to be contrasted to the 'targeting' approach adopted by the Court.[40]

The dis-targeting approach obligates businesses actively to regulate which jurisdictions they serve. This is clearly a rather heavy burden on businesses. However, I am hopeful that whatever burden this presents would be outweighed by the greater degree of predictability. Furthermore, businesses must be provided with adequate possibilities to take reasonable steps that rebut the presumption that they are targeting the world at large. Businesses must be furnished with the ability to take active, yet simple, steps to avoid the risk of entering into contracts with consumers in 'undesirable' forums; this is an essential component of the dis-targeting approach. What can be

[39] I acknowledge that this assumption will be severely undermined if the EU adopts the Proposal for a Regulation of the European Parliament and of the Council on addressing geo-blocking and other forms of discrimination based on customers' nationality, place of residence, or place of establishment within the internal market and amending Regulation (EC) No 2006/2004 and Directive 2009/22/EC, COM(2016) 289 final.

[40] For a discussion of the advantages of dis-targeting, see eg Dan Svantesson, 'Pammer and Hotel Alpenhof—ECJ Decision Creates Further Uncertainty About When E-Businesses "Direct Activities" to a Consumer's State under the Brussels I Regulation' (June 2011) 27(3) Compu L & Sec Rev 298–304, 302–04. However, I first put forward the idea in Dan Svantesson, 'What Should Article 7—Consumer Contracts, of the Proposed Hague Convention, Aim to Accomplish in Relation to E-Commerce?' (September/October 2001) 17(5) Compu L & Sec Rep 318–25.

regarded as reasonable steps must be viewed in light of technological and legal developments.[41]

One of the problems is that the targeting approach, not least as applied by the Advocate General and the Court in *Pammer/Hotel Alpenhof*, is misguided in its focuses on the subjective intentions of the relevant party. Both the Advocate General and the ECJ seem to have taken the view that the phrase 'directs such activities' implies a conscious decision executed without mistakes. The Advocate General states: 'It is therefore essential for there to be active conduct on the part of the undertaking, the *objective and outcome* of which is to win customers from other Member States.'[42]

While I agree that there must be some active conduct on the part of the undertaking, the problem with this approach is found in the words in italic. We can have a situation in which the objective of an activity is to win customers from other Member States, but no such outcome is realised. We can also have a situation in which the outcome of an activity is to win customers from other Member States, without it being the trader's objective. The better approach is to distinguish between *objective* and *outcome* and only focus on the outcome when determining whether the business has directed its activities to the consumer's state. Such a move from subjectivity to objectivity promotes certainty and fairness.

At any rate, the targeting approach has a rather long history—as is illustrated in cases such as *Ward Group Pty Ltd v Brodie & Stone plc* of 2005.[43] In fact, it may be appropriate to view it as the offspring of the 'effects doctrine' known from public international law. Nevertheless, since the *Pammer/Alpenhof* decision the targeting approach has, particularly in European literature and reasoning, assumed a most elevated position; it is in this principle some, predominantly Europeans, seem to be placing their faith. And, indeed, the targeting approach has made its way into fields outside the consumer protection area and is now also

[41] I note that if the EU adopts the Proposal for a Regulation of the European Parliament and of the Council on addressing geo-blocking and other forms of discrimination based on customers' nationality, place of residence, or place of establishment within the internal market and amending Regulation (EC) No 2006/2004 and Directive 2009/22/EC, COM(2016) 289 final, the option of the use of geo-location technologies as a 'reasonable step' would be removed, which would be a serious—if not fatal—issue for my proposed approach.

[42] Advocate General Trstenjak 2010, *Opinion of Advocate General Trstenjak delivered on 18 May, Case C-585/08* Peter Pammer v Reederei Karl Schlüter GmbH & Co KG *and Case C-144/09* Hotel Alpenhof GesmbH v Oliver Heller, 18 May <http://curia.europa.eu/juris/document/document_print.jsf?doclang=EN&docid=79076&cid=594463> [63] (emphasis added), accessed June 2017.

[43] *Ward Group Pty Ltd v Brodie & Stone plc* (with Corrigendum dated 19 May 2005) [2005] FCA 471.

the 'weapon of choice' for the EU's latest attempt at dealing with data protection.[44]

However, much is wrong with the overly simplistic targeting approach and elsewhere[45] I have described it as the legislator's dream but the judge's (and indeed lawyer's) nightmare. Although it in theory comes across as providing certainty, predictability, and simplicity, it is hopelessly difficult to apply in practical situations;[46] it will provide no certainty for the parties involved as courts frequently are going to have to conclude either that they target just about every country in the world or no countries at all.[47] Both these options will render the targeting approach useless, which undermines the legitimacy of the instruments that rely upon it. Moreover we do well to remember that not only is exorbitant jurisdiction a problem, but arbitrary jurisdiction is also a serious concern.

Another important component of the emerging under-regulation is exemplified in the US' adoption on 10 August 2010 of a Federal statute seeking to address what the US perceives as 'libel tourism'.[48] The key feature of the statute is to make mandatory the non-recognition of foreign defamation judgments that are seen as inconsistent with the First Amendment's protection of free speech. With actual enforcement being such a central component in the proper functioning of the private international law machinery, the approach taken by the US is doubtless a step in the direction of under-regulation. The obvious risk is that this initiative prompts the response that other countries implement similar non-recognition legislation as to areas of law in which

[44] Regulation (EU) 2016/679 of the European Parliament and of the Council of 27 April 2016 on the protection of natural persons with regard to the processing of personal data and on the free movement of such data, and repealing Directive 95/46/EC (General Data Protection Regulation), [2016] OJ L 119/1–88, Art 3.

[45] Dan Svantesson, 'Extraterritoriality and Targeting in EU Data Privacy Law: The Weak Spot Undermining the Regulation' (2015) 5(4) Intl Data Priv L 226–34.

[46] See also Maja Brkan, 'The Unstoppable Expansion of EU Fundamental Right to Data Protection. Little Shop of Horrors?' (2016) 23(5) Maastricht Journal of European and Comparative Law 812–41, 834–37, as well as Oster noting that it is questionable whether the targeting doctrine can be properly applied to many different types of tort (Jan Oster, *European and International Media Law* (CUP 2017) 134).

[47] These issues would be all the more serious if the EU adopts its Proposal for a Regulation of the European Parliament and of the Council on addressing geo-blocking and other forms of discrimination based on customers' nationality, place of residence, or place of establishment within the internal market and amending Regulation (EC) No 2006/2004 and Directive 2009/22/EC, COM(2016) 289 final, since doing so would remove the possibility of relying on the presence, or absence, of geo-location technologies as an indicator of targeting.

[48] See further EC Barbour, *The SPEECH Act: The Federal Response to 'Libel Tourism'* (Congressional Research Service, 2010) <http://www.fas.org/sgp/crs/misc/R41417.pdf>, accessed 3 April 2017.

they do not trust how US courts adjudicate matters. In other words, the US initiative promotes a downward spiral effect.

It is also noteworthy that, during this period, we began to see an increase in technology-specific, as opposed to technology-neutral, lawmaking. The most famous example can be found in the CJEU's decision in the joined cases of *eDate Advertising GmbH v X*[49] and *Olivier Martinez, Robert Martinez v MGN Ltd*,[50] where the Court dealt with online publications differently to offline publications. Other examples can also be found.[51]

As to the academic literature of this (comparatively shorter) period, I think it is fair to say that, sometime during this period, Internet jurisdiction became a mainstream issue. This is, for example, reflected in the wealth of journal articles published on the topic[52] and in the fact that we saw Internet jurisdiction-specific events being organised.[53] Another indication of the status Internet jurisdiction matters gained during this period is found in the various major studies released such as that of the Max Planck Group on Conflict of Laws in Intellectual Property[54] and the comprehensive and detailed report 'Internet Jurisdiction: A Survey of Legal Scholarship Published in English and United States Case Law' published by the Center on Law and Information Policy at the Fordham University School of Law in 2013.[55] A second report was published by the Fordham University School of

[49] [2011] Referring court Bundesgerichtshof Germany Case C-509/09.

[50] [2011] Referring court Tribunal de grande instance de Paris France Case C-161/10.

[51] To name but one example, on 28 October 2010 the Standing Committee of China's National People's Congress adopted the *Law of the People's Republic of China on the Application of Law for Foreign-Related Civil Relations 2011*. Article 46 deals specifically with Internet defamation: 'Where such personal rights as the right of name, portrait, reputation and privacy are infringed upon via network or by other means, the laws at the habitual residence of the infringed shall apply.'

[52] I will not engage here in anything as foolish as an attempt to map out all the relevant literature. However, some publications stand out, such as Kuner's two-part journal article on jurisdiction in data privacy law (Christopher Kuner, 'Data Protection Law and International Jurisdiction on the Internet (Part 1)' (2010) 18(2) Intl JL & Info Tech 176; Christopher Kuner, 'Data Protection Law and International Jurisdiction on the Internet (Part 2)' (2010) 18(3) Intl JL & Info Tech 227); Trimble's work on geo-location (eg M Trimble, 'The Future of Cybertravel: Legal Implications of the Evasion of Geolocation' (2012) 22 Fordham Int Prop, Media & Ent LJ 567); Chris Reed, *Making Laws for Cyberspace* (OUP 2012).

[53] Such as 'Internet Jurisdiction Symposium: About the Symposium' <http://www.aber.ac.uk/en/internet-jurisdiction/about/>, accessed 4 April 2017.

[54] European Max Planck Group on Conflict of Laws in Intellectual Property, *Conflict of Laws in Intellectual Property: The CLIP Principles and Commentary* (OUP 2013).

[55] Joel R Reidenberg et al, 'Internet Jurisdiction: A Survey of Legal Scholarship Published in English and United States Case Law' (30 June 2013) Fordham Law Legal Studies Research Paper No 2309526 <http://ssrn.com/abstract=2309526> or <http://dx.doi.org/10.2139/ssrn.2309526>, accessed 4 April 2017.

Law, dealing specifically with German scholarship and cases.[56] Scholarship such as this is of great value in mapping out trends—trends that then can be analysed and structured.

Another initiative that was born during this period is the Internet & Jurisdiction policy network, whose Secretariat is under the Directorship of Mr de La Chapelle: 'Internet & Jurisdiction is the global multistakeholder policy network addressing the tension between the cross-border Internet and national jurisdictions. It facilitates a global policy process to enable transnational cooperation and preserve the global character of the Internet.'[57]

4.5 An Era of Hyper-regulation (2015–?)

During the current era, Internet jurisdiction remains amongst the 'hot' topics dealt with at various conferences, workshops, etc. In addition, there is a steady flow—not to say flood—of writings on the topic. Here, I want to pause to mention that, while some commentators (myself included) have written on Internet jurisdiction for many years and continue to do so, I am particularly hopeful that the numerous younger scholars now working on their doctoral dissertations, journal articles, book chapters, or books on this topic will make more progress than we have managed to achieve so far. We are in desperate need of fresh perspectives.

Important work is also being done, specifically on Internet jurisdiction, by key organisations. To mention but two examples, the Council of Europe is undertaking significant reform work that is likely to expand the role of the Budapest Convention,[58] and also the EU is working on jurisdictional issues in the context of the Internet, and Internet crime.[59]

On many occasions, both during speaking engagements and in writings, I have used the same example to illustrate the fact that we are now in an era of hyper-regulation. The example I use is focused on the number of laws we

[56] Desiree Jaeger-Fine et al, 'Internet Jurisdiction: A Survey of German Scholarship and Cases' [30 June 2013] Fordham Law Legal Studies Research Paper No. 2309575 <http://ssrn.com/abstract=2309575> or <http://dx.doi.org/10.2139/ssrn.2309575>, both accessed 4 April 2017.

[57] 'Internet & Jurisdiction Policy Network: Mission', <http://www.internetjurisdiction.net/about/mission>, accessed 4 April 2017. For the 2015/2016 Progress Report, see <http://www.internetjurisdiction.net/uploads/pdfs/Annual-Reports/Internet-Jurisdiction-2015-16-Report.pdf>, accessed 8 April 2017.

[58] Council of Europe Convention on Cybercrime (ETS No 185). See further: <http://www.coe.int/en/web/cybercrime/ceg>, accessed 11 April 2017.

[59] Council conclusions on improving criminal justice in Cyberspace (Luxembourg, 9 June 2016) <https://ec.europa.eu/home-affairs/sites/homeaffairs/files/what-we-do/policies/organized-crime-and-human-trafficking/council_conclusions_on_improving_criminal_justice_in_cyberspace_en.pdf>, accessed 11 April 2017.

are exposed to when engaging in a mundane activity such as posting something on social media. What laws do we need to consider in doing so?

In most instances, it seems beyond intelligent dispute that you will have to take account of the law of the state you are in at the time you make the posting. However, that is, of course, not the end of the matter. You may also need to consider the law of the state in which you are habitually residing (and/or domicile) and the law of your state of citizenship. Then you will probably also need to consider US law, as most major social media platforms are based in the US.[60] I am here already talking of a few, potentially very different, legal systems supplying laws you are meant to comply with.

If your posting relates to another person, you may also need to consider the laws of that person's location, residence, domicile, and citizenship. You may also need to consider the laws of any additional state in which that person has a reputation to protect.

But then, under the law of many, not to say most, states focus may be placed on where content is downloaded or read. This means that you will also need to take account of the laws of all the states in which your Facebook 'friends' or LinkedIn 'connections' are found; and less predictably, the laws of all the states in which they may be located when reading your posting. It goes without saying that the number of additional legal systems to be considered grows with the number, and geographical diversity, of your friends or connections, and in the light of the mobility of people, may never be fully ascertained at the time of posting.

Then things get rather messy. Given that your postings may be re-posted, you also need to take account of the laws of all the states in which re-posted versions of your posting may be downloaded or read. Here the original poster obviously loses all possibility of predicting the scope of laws to which she or he may be exposed.

As if this was not complicated enough, we must also bear in mind that content placed on social media platforms is often stored in 'the cloud', and while we as users may not necessarily be able to find out where our content is stored, we may be legally obligated to consider the laws of the state in which it is stored. Finally, content posted may, depending on both your settings and on how your social media platform treats those settings, be available to third parties and you may then need also to let the laws of the locations of those third parties guide your conduct.

[60] While the validity of online choice of forum and choice of law clauses has been called into question to a degree through recent decisions such as the CJEU's judgment in C-191/15 (2 June 2016) and the Supreme Court of Canada's decision in *Douez v Facebook, Inc* 2017 SCC 33 (23 June 2017), it remains likely that such choices will be upheld in many cases.

This legal situation, of extraordinary complexity, is what 2 billion Facebook users expose themselves to on a daily basis. And of course, a similar reasoning could be applied to users of other social media platforms such as LinkedIn and Google+.

For lawyers, this brings attention to something important. While the conventional way in which we approach the choice of law questions is to ask which state's law is the applicable law, the online environment makes such an approach obsolete; it is clearly not a matter of identifying the applicable law, it is a matter of assessing which of many applicable laws actually matter. In other words, it is a pure risk assessment exercise. Examining the implications of this change in detail would, however, take us too far afield, and this topic will thus not be pursued further here.

But what are then the implications for the ordinary Internet users? For the absolute majority, their postings will not lead to any legal drama. However, the thought of being exposed to potential legal liability in a large number of states should be a concern to anyone. And of course, the very idea that strictly speaking you should inform yourself of all those laws you are meant to follow is daunting indeed. This may be worth considering next time you are posting something on your favourite social media platform.

Of course, as noted for example by Crawford: 'There is no assumption (even in criminal cases) that individuals or corporations will be regulated only once, and situations of multiple jurisdictional competence occur frequently.'[61] In other words, we have no legitimate expectation of being governed by just one state's laws. However, the hyper-regulation to which we are exposed online is unique to the online environment; a fact we must not overlook when discussing Internet jurisdiction.

Discussing the consequences of hyper-regulation, we may usefully recall Fuller's famous eight 'distinct routes to disaster' in lawmaking which he used to highlight an 'inner morality of law':

> The first and most obvious lies in a failure to achieve rules at all, so that every issue must be decided on an ad hoc basis. The other routes are: (2) a failure to publicize, or at least to make available to the affected party, the rules he is expected to observe; (3) the abuse of retroactive legislation, which not only cannot itself guide action, but undercuts the integrity of rules prospective in effect, since it puts them under the threat of retrospective change; (4) a failure to make rules understandable; (5) the enactment of contradictory rules or (6) rules that require conduct beyond the powers of the affected party; (7) introducing such frequent changes in the rules that the

[61] James Crawford, *Brownlie's Principles of Public International Law* (8th edn, OUP 2012) 457.

subject cannot orient his action by them; and finally, (8) a failure of congruence between the rules as announced and their actual administration.[62]

If we accept Fuller's framework as the benchmark, the consequence of our current degree of hyper-regulation is nothing less than the complete bankruptcy of the legal system as applied to the online environment. Our example must have made clear that a person making a posting on social media typically is unable to know what states' laws she is expected to abide by; this is so, as we need to take account of factors such as where readers of re-posted versions of our postings are located when reading the postings—factors we cannot control or predict.

But even if we miraculously managed to identify all the states whose laws we need to consider, how would we access those laws? For some states, we may rely on free legal databases such as the excellent resources provided by the Australasian Legal Information Institute (AustLII). However, many states lack such databases. And even where such databases are available, they may be in a language we do not understand;[63] and only the brave and the foolish would want to rely on currently available automated translations for something as complex as a legal text. Thus, it is questionable that the situation we are in survives Fuller's second and fourth routes to disaster in law-making: hyper-regulation involves a 'failure to publicize, or at least to make available to the affected party, the rules he is expected to observe',[64] and it represents a 'a failure to make rules understandable'.[65]

Perhaps it may be said that we have made remarkably little progress on the important topic of authorised translations of laws. After all, as far back as the nineteenth century Bentham called for such translations, observing that they:

would prevent a stranger from falling into those faults which he might otherwise commit through ignorance of the law, and also guard him from the snares which otherwise might be laid for him by abusing his ignorance. Hence would arise security for commerce, and confidence in transactions among foreign nations. It is a proceeding called for by candour and honesty.[66]

[62] Lon L Fuller, *The Morality of Law* (2nd edn, Yale University Press 1969) 39.

[63] I am here referring to the fact that the relevant legal texts are likely to be presented in a foreign language. To this we may, of course, add the complication stemming from the lawyers' habit of expressing themselves in what has been called 'legalese' (see further Ewoud Hondius, 'The Unbearable Heaviness of Legal Texts' in Børge Dahl, Thomas Riis, and Jan Trzaskowski (eds), *Liber Amicorum Peter Møgelvang-Hansen* (Ex Tuto 2016) 141–48). This practice exacerbates the problem.

[64] Fuller (n 62) 39. [65] ibid.

[66] Jeremy Bentham, *Of Promulgation of the Laws and Promulgation of the Reasons Thereof* (1843) <http://www.laits.utexas.edu/poltheory/bentham/promul/index.html>, accessed 4 April 2017.

Furthermore, the costs of being globally legally aware are astronomical:

> Most individuals have only a general impression of the rules of their own national law, and are likely to be completely ignorant of the multiplicity of foreign laws which claim to apply to their cyberspace activities. This is even true for online businesses with customers in multiple jurisdictions. The costs of discovering the potentially applicable laws of over 190 states, and then analysing their potential application to the business's activities, are so great that such businesses tend only to investigate jurisdictions where they perceive themselves as potentially at risk.[67]

Finally, if all the aforementioned obstacles were overcome—and I cannot see how they could be—what chance does the average citizen have of understanding the exact operation of complex, and sometimes contradictory, legal texts, cases, and other documents from a diverse range of different states? Texts, cases, and other documents not even trained lawyers and judges can agree upon the meaning of.

Here, we may go beyond Fuller's framework and connect with the concept of 'the rule of law'. As noted by Ireland-Piper, the rule of law is a contested concept.[68] Yet, one of the bare minimum requirements it incorporates is that the law must be readily knowable, certain, and clear.[69] As has been demonstrated, the jurisdictional claims of our era of hyper-regulation are not knowable, nor are they certain or clear.

The question then is how the complications identified here can be reconciled with the principle that ignorance of the law is not a defence.[70] This principle is a fundamental feature of all legal systems and dates back at least to Roman law. Its logic, as well as its necessity, is easily recognised; if ignorance of the law would be a defence, there would be an incentive to remain ignorant of the law, and it would prove difficult indeed to succeed in any legal action. In the light of this, while some exceptions to a strict application of this principle can be found,[71] I suspect it would be futile to argue against it in the context of domestic legal systems. In fact, a removal of this principle may arguably be seen to herald the collapse of the legal system. However, at the same time, given the above, can the principle that ignorance of the law is no excuse really survive online? Surely the law also needs to re-evaluate its cornerstones from time to time; and where a cornerstone fits so poorly with reality that no one can take it seriously, is it not then the law's duty to adjust?

[67] Reed (n 52) 13–14 (internal footnotes omitted).
[68] Danielle Ireland-Piper, *Accountability in Extraterritoriality* (Edward Elgar 2017) 5.
[69] ibid 6.
[70] *Ignorantia juris non excusat, ignorantia legis neminem excusat* or *nemo censetur ignorare legem*.
[71] See eg The American Law Institute's Model Penal Code, Section 2.04, Ignorance or Mistake.

Be that as it may, regardless of the acceptance of the principle that ignorance of the law is not a defence in domestic law, we must question its legitimacy where domestic law spills over into other states: where one state gives its laws an excessively wide scope of application.

The conclusion must be that the legal fiction of the legally aware citizen—necessary as it may be—is severely out of sync with reality, at least when it comes to the 'extraterritorial' application of domestic laws to Internet conduct. As observed by Reed: 'Ignorance of foreign law is not just common in cyberspace; it is inevitable.'[72]

Earlier we approached the principle that ignorance of the law is not a defence from the perspective of the people's awareness of the existence of laws. However, laws are often highly technical and rarely accessible to the general public—even where people are aware of a particular legal statute, for example, they may lack the ability to understand what the law actually demands of them. Here again we may connect to Fuller and his fourth distinct route to disaster in lawmaking: 'a failure to make rules understandable'.[73] Importantly, Fuller points out that '[t]he need for this education [the education of citizens as to the content of the laws that apply to them] will, of course, depend upon how far the requirements of law depart from generally shared views of right and wrong.'[74] It is easy to agree with this, and the implications it has for the current discussion could escape no one. The laws we are exposed to when acting online are diverse and come from virtually all domestic legal systems in the world. The fact that those rules will then include concepts and principles that are different to the concepts and principles of right and wrong generally shared within our respective communities is beyond intelligent dispute.

The idea that education as to the content of the laws is only needed where those laws depart from generally shared views of right and wrong has a certain first-glance appeal. Thus, it may be tempting to conclude that in this we have found the real core of the issue: that, as long as laws conform to generally shared views of right and wrong, other matters like those laws being publicised and drafted in clear, understandable language are of mere subordinate importance. Tempting as it may be, such a conclusion lacks merit. After all, our laws do not only regulate matters that may be referred to generally shared views of right and wrong. Whether we should pay 35 or 36 per cent tax can hardly be decided based on generally shared views of right and wrong. Similarly, such generally shared views cannot tell us whether a consumer should enjoy 14 or 15 days' right to return products bought through

[72] Reed (n 52) 71. [73] Fuller (n 62) 39. [74] ibid 50.

a distant purchase. Neither of these decisions depends on societal norms of right and wrong—they are not norm-based lawmaking decisions. Thus, laws expressed in clear, understandable language being publicised remains of fundamental importance.

The observation that not all lawmaking decisions are linkable to societal norms is important in the context of current discussions about how to make laws for the Internet. For example, a key argument in Reed's interesting *Making Laws for Cyberspace*[75] is, it seems to me, that at least those laws that are difficult to enforce must be in line with Cyberspace norms to be obeyed.[76] Perhaps it may be argued that Reed then does not give sufficient attention to the role the law plays in drawing more or less arbitrary lines in *norm-neutral matters* such as the question of a 35 or 36 per cent tax rate? After all, the making of laws for Cyberspace also involves such line-drawing in norm-neutral matters.

Returning to the matter of the importance of legally informed citizens, I hasten to acknowledge that it is utopian to call for a situation where every person is informed of every rule they are expected to follow in every legal system to which they are exposed. Indeed, Fuller goes as far as to conclude that '[i]t would in fact be foolish to try to educate every citizen into the full meaning of every law that might conceivably be applied to him.'[77] However, the fact remains that the global reach of Internet communications has placed the legal system on a route to disaster for the average Internet user. Even if it was the case, and it is not, that states made all their laws available online, language barriers and difficulties in locating the laws (not to mention understanding them) will defeat even the Internet user keenest to abide by all applicable laws. While lacking wicked intent, the situation is somewhat akin to Caligula respecting the tradition of Roman laws being posted in a public place but ensuring that his laws 'were in such fine print and hung so high that no one could read them'.[78] Unfortunately, this fact has been widely recognised yet ignored for too long.

4.6 Concluding Remarks as to the History of Internet Jurisdiction

This chapter provided a brief history of Internet jurisdiction, taking account of key court decisions and legislation as well as developments in the academic thinking on the topic. In doing so, it divided the history of Internet

[75] Reed (n 52). [76] See eg ibid 220. [77] Fuller (n 62) 49. [78] ibid 49.

jurisdiction into four relatively distinct phases, and briefly discussed the pre-1991 landscape.

This discussion lends itself to several conclusions. First of all, it is clear that law has largely been reactive, responding to technological developments. Second, the level of creativity applied in the search for workable solutions was seemingly higher in the earlier stages than in more recent times. Third, the attitudes of courts, legislators, and the academic community have varied considerably over the time period examined. Fourth, academic opinions remain divided. Fifth, many different solutions have been proposed, none of them gaining global (or near global) acceptance. Sixth, there is a strong and consistent recognition and awareness of the problems that arise in the context of Internet jurisdiction. Seventh, there is a widespread dissatisfaction with conventional jurisdictional doctrines and thinking. Eighth, international global consensus seems unlikely any time in the foreseeable future. And finally, ninth, the current landscape of hyper-regulation is clearly unsatisfactory, and indeed, it is arguable that the situation we are in now in is the worst of all the different phases discussed—in other words, while our understanding of the complex issues involved is increasing, it is questionable whether we are making progress with the actual problems involved.

5

Jurisdictional Interoperability—the Path Forward (for Now)

The previous chapters have made clear that, on the topic of Internet jurisdiction, the difference between how things are and how things ought to be is great indeed. Perhaps the strongest and most obvious lesson we can learn from the history of Internet jurisdiction is that we are not in a position to reach global (or even widespread) consensus on the area of Internet jurisdiction as a whole. To this, we may add that the prospect of global (or even widespread) consensus on all areas of substantive law is, if anything, even slimmer. While widespread consensus is achievable in some rare areas (such as eg child abuse materials), most areas are clearly nowhere near being ready for global, or near global, harmonisation.

The underlying differences in cultural and societal values, as well as in attitudes towards Internet jurisdiction, are strong at the present. So are the methodological differences between various legal systems.[1] And due to the many and important differences in substantive law, states have a strong interest in pursuing the application of their respective laws in matters that affect them, their interest, their territories, or their citizens. Indeed, this is by no means a novel concern. Consider, for example, this statement from Nyholm's dissenting opinion in the *Lotus* case:

> As a method of regulating the relations between countries, in the first place should be mentioned more or less universal conventions concluded between States and serving to bridge over the domain not subject to any regulation. Universal laws adopted by all countries and having as their object the creation or the codification of international law would constitute a solution of the problem, but they do not exist and one can only endeavour to establish international law by custom.[2]

[1] See eg Jan Oster, 'Which Limits on Freedom of Expression are Legitimate? Divergence of Free Speech Values in Europe and the United States' in Uta Kohl (ed), *The Net and the Nation State—Multidisciplinary Perspectives on Internet Governance* (CUP 2017) 39–47, 46–47.

[2] SS 'Lotus' (*France v Turkey*) (1927) PCIJ Series A, No 10, 58.

Solving the Internet Jurisdiction Puzzle. First Edition. Dan Jerker B. Svantesson.

Here it is interesting to ponder the following. Imagine that consensus on a uniform universally applicable legal system could be reached. Imagine further that a person in State A posts content online that defames a person in State B. Obviously, the choice of law issue is easily settled—after all there is only one legal system and it applies in both State A and in State B. However, that alone does not ensure a uniform application of the law; after all, how laws are interpreted varies between countries and depends on various cultural and societal factors. Thus, even if we have the exact same law applicable in both State A and State B, the courts in the respective countries may interpret the law differently, which means that Internet jurisdiction still matters despite the introduction of a uniform universally applicable legal system. Moreover, if someone wishes to object that, by talking of our fictional uniform universally applicable legal system, we ought to mean a legal system that is uniform not just in wording but also in interpretation and application, the fact remains that people nevertheless may prefer to litigate in one court over another due to reasons such as convenience, language, or greater familiarity with the legal profession in the preferred country. In the light of this, the introduction of a uniform universally applicable legal system remains firmly established as an impossible solution to the Internet jurisdiction puzzle.

Clearly, then, our conundrum is associated with the choice of forum issue as well as the choice of law issue. And here, I suggest, we have found the perhaps disappointingly simple core of the Internet jurisdiction puzzle—it arises as a consequence of four factors:

1. The Internet lends itself to cross-border communication;
2. the Internet, and Internet activities, must be subject to the law;
3. where disputes arise, it matters which law is applicable; and
4. where disputes arise, it matters which court (or other body) determines the dispute.

To this we may add the problems stemming from the previously mentioned glacial speed with which law develops. Just consider the way the Internet developed during the time the Hague Conference's 'Judgments Project 1.0' was being negotiated. Viewing the 2005 Convention on Choice of Court Agreements[3] as a result of the Judgments Project 1.0, we can note how, in the thirteen years it took for the Convention to be finalised, e-commerce and Internet usage went from its infancy to boom to burst, then into recovery resulting in the blossoming and multifaceted

[3] Hague Convention of 30 June 2005 on Choice of Court Agreements.

communications medium we have today. Expressed in numbers, the development is quite remarkable. In 1992, there were approximately 1,000,000 Internet hosts, compared with 400,000,000 in 2005.[4] Furthermore, there were only about 50 websites on the Internet in 1992, compared with over 100,000,000 in 2005.[5] The dramatic change highlighted by these figures makes clear that, in drafting complex international agreements, the drafters must necessarily adopt a degree of technology-neutral language. It would, for example, have been unsuitable for the drafters to include provisions dealing with bulletin board systems, as the function of such systems has been largely replaced by websites, blogs, and social networking sites.[6]

Yet perhaps the most commonly proposed solution to the Internet jurisdiction problems is harmonisation via international agreement. Most commentators (myself included) have at some stage reached the conclusion that the best, if not only, path forward is to work towards an international agreement addressing the jurisdictional complications that plague the online environment. While resorting to such a conclusion reasonably may be added to the list of *peccata mortalia*, such conclusions are not reached out of naivety; most commentators stress the considerable obstacles that stand in the path of this 'Holy Grail'. However, the reality is unfortunately that if this is the best proposal we manage to come up with, we are failing. Indeed, such suggestions, due to their slim prospects of success, are simply not good enough.

We must abandon hope of giant leaps taking us to our goal and accept instead that the road ahead will be travelled by many thousands of small steps. Just as the Internet is a successful network of networks, the mid-term solution to the jurisdictional issues online will be found in what we can see as a *legal system of systems*—a system in which our domestic legal systems operate smoothly together with a minimum of inconsistencies and clashes.

Currently, the interaction between different domestic legal systems seeking to regulate our conduct online does not work smoothly; as can be expected, there are considerable inconsistencies and, indeed, outright clashes. On a state level, this takes the form of competing jurisdictional claims, and indeed, in certain situations that may be seen as gaps in jurisdictional claims, that is, where no state claims jurisdiction. However, for persons—whether

[4] Robert H Zakon, 'Hobbes' Internet Timeline' <http://www.zakon.org/robert/internet/timeline/>, accessed 4 April 2017.

[5] ibid.

[6] Dan Svantesson, 'The Choice of Courts Convention: How Will It Work in Relation to the Internet and E-commerce?' (2009) 5(3) J Priv Intl L 517–35, 519.

legal or natural—these inconsistencies and clashes are arguably pronounced in an even more complex manner.

Rights given under one legal system frequently clash with duties imposed by other legal systems. And occasionally the duties imposed by one legal system clash with the duties imposed by another legal system, leaving the subjects with the precarious decision of which legal system's duties to ignore. From the example earlier of the number of laws to which a social media user is exposed, it is clear that Internet users are exposed to a complex matrix of legal systems. Thus, to speak of their legal system in a meaningful manner, I have suggested that we need to introduce the concept of a 'contextual legal system', by which I refer to the system of legal rules from all sources that purport to apply to the conduct of the person in question in the setting in which he or she is acting.[7] When we choose to drive over the speed limit, our contextual legal system is typically that of the country in which we are driving, and if we, in country A, post a letter to someone in country B defaming someone in country C, the contextual legal system may be made up of the relevant aspects of the laws of at least countries A, B, and C.

Where this reasoning is accepted, it is clear that the contextual legal system Internet users face is made up of parts of the laws of many different countries. It should therefore surprise no one that our contextual legal system in such a situation will typically contain legal rules overlapping and clashing with other legal rules. Indeed, in such a contextual legal system, we are likely to see instances of legal rules directly contradicting each other, leaving the Internet user with the perilous exercise of selecting which law to comply with and which to ignore.

To make our domestic legal systems operate smoothly together with a minimum of inconsistencies and clashes our aim should be to create what I have elsewhere[8] termed *jurisdictional interoperability* between the different domestic legal systems to the greatest degree possible. No doubt, aiming for smooth perfection is to set the bar too high. However, by identifying any uniting features (of which there are many), and in seeking to iron out inconsistencies and clashes between domestic legal systems, both in substantive and procedural rules, much can be achieved.

[7] Dan Svantesson, 'The Holy Trinity of Legal Fictions Undermining the Application of Law to the Global Internet' (2015) 23(3) Intl JL & Info Tech 219–34.

[8] ibid.

5.1 Jurisdictional Interoperability as a Part of Legal Interoperability

The term 'legal interoperability' is often used in the context of research data. However, it is its use in a somewhat different setting that concerns us here. In their interesting, and indeed ground-breaking, 2012 book *Interop: The Promise and Perils of Highly Interconnected Systems*, Palfrey and Gasser discuss legal interoperability in terms of 'the process of making legal norms work together across jurisdictions'.[9] In their view, there are three reasons why we need a higher degree of legal and policy interoperability:

> At the most basic level, legal interoperability can reduce the costs associated with doing business across borders. […] More broadly, recent studies suggest that legal interoperability—particularly in the fields of trade and business law—drives innovation and economic growth by making countries more competitive. Finally, increased levels of legal interoperability among multiple jurisdictions can lead to better laws that foster the development of fundamental values and rights, such as information privacy and freedom of expression.[10]

Discussing legal interoperability as a normative tool, Weber draws upon the work of Palfrey and Gasser and makes the point that:

> [t]he supranational realization of the process of legal interoperability (as the process of making legal rules work together across jurisdictions) can fluctuate between two poles: full harmonization and a complete fragmentation on a bilateral, plurilateral or multilateral level. In the first scenario, all laws would be the same everywhere; in the second, the legal systems would be so different in each country that economic, social and cultural interactions become impossible […].[11]

To me, however, complete legal interoperability cannot depend—at least in theory—on all laws being the same everywhere. Rather, complete legal interoperability is achieved where there are no clashes between any laws anywhere: that is, where all rights and obligations stemming from all legal systems, that are part of the system in question, exist in complete harmony.[12]

[9] John Palfrey and Urs Gasser, *Interop: The Promise and Perils of Highly Interconnected Systems* (Basic Books 2012) 178.

[10] ibid 178–79.

[11] RH Weber, 'Legal Interoperability as a Tool for Combating Fragmentation' (2014) Global Commission on Internet Governance Paper Series No 4 (Centre for International Governance Innovation) 6.

[12] Perhaps it is this Weber refers to when making the statement that '[l]egal interoperability should make systems work together, but not make the systems all the same' (ibid 6)—a statement

Given this understanding, legal interoperability is the standard goal of virtually every domestic legal system in that they all strive towards ensuring that there are no norms that contradict other norms and, at the minimum, where such clashes arise, the legal system provides a mechanism for deciding which norm trumps the other conflicting norm. Indeed, as Fuller stresses, the enactment of contradictory rules is one of the distinct routes to disaster in lawmaking.

It is only logical to extend this ambition, so typical of domestic legal systems, to the international context, and where we do so, it is the jurisdictional rules of public and private international law that, in a sense, perform the role of deciding which norms trump the other conflicting norms. This is an aspect of legal interoperability focused on procedural law. However, as procedural law contains many elements going beyond jurisdictional matters, it is useful to distinguish it both from the broader term of procedural legal interoperability and from the even broader concept of legal interoperability as such. It is against this background that I discuss this topic as 'jurisdictional interoperability'.

Weber apparently attributes rather limited value to what I here have described as jurisdictional interoperability:

> [T]he rules on conflicts of law (private international law) do not overcome the substantive differences in national legal orders (and therefore do not lead to legal interoperability), they only give guidance on how to determine the applicable normative rules. This assessment does not mean that such rules do not have any impact on the substantive contents of interoperable legal systems, but their influence is of an indirect nature.[13]

In response to this, I want to make two observations. First of all, if we had a flawlessly functioning system for allocating jurisdiction and for deciding the applicable law, it would not matter that there were substantive differences in national legal orders. After all, in this utopian scenario, those substantive differences in national legal orders could never clash as our flawlessly functioning system for allocating jurisdictional and deciding the applicable law on every occasion would prevent such clashes. Secondly, even if we do not resort to such a wondrous fairy-tale landscape—relevant as it may be for a theoretical discussion—and return to the harsh reality, I argue that jurisdictional interoperability probably has more to offer than does legal interoperability focused on substantive law. To use but one example, and indeed an

that surely contradicts his assertion that complete legal interoperability arises where 'all laws would be the same everywhere'.

[13] ibid.

example Weber himself discusses, 'the likelihood of the First Amendment to the United States Constitution (which includes the freedoms of religion, speech, the press and association) becoming the rule in China or the Middle East is extremely low'.[14] In other words, I think that, for many areas of law, we have a better prospect of achieving legal interoperability in the form of when a state's law applies than we do when it comes to achieving legal interoperability in the form of what that state's law should be. I hasten to acknowledge, however, that this is context-dependent.

5.2 Features of Jurisdictional Interoperability

Given that what I here discuss as jurisdictional interoperability is nothing but a particular aspect of legal interoperability, it is only natural that, in a pursuit of jurisdictional interoperability, we can learn a lot from Palfrey and Gasser's observations regarding legal interoperability:

1. We need not pursue 'the goal to create one uniform "world law"'.[15]
2. 'We need to aim for interoperability among legal systems at an optimal, rather than maximum, level'.[16]
3. Legal interoperability is of significance for a diverse group of actors including '[l]awmakers and policy makers, as well as businesspeople'.[17]
4. '[L]egal interoperability is a matter of degrees rather than an all-or-nothing prospect.'[18]
5. 'Higher or lower degrees of legal interoperability may be appropriate in different circumstances.'[19]
6. 'Even where "law on the books" is roughly harmonized, there is often a great deal of variability in terms of how it is enforced across countries and courts. The net result is a lower level of interoperability as far as "law in action" is concerned.'[20]
7. 'If the proper balance is struck, states can maintain a strong degree of local decision making over aspects of the law that matter to them, while creating ways for people and businesses to work together across national borders.'[21]
8. 'Legal interop is not a matter of set-and-forget; it requires ongoing diligence and care.'[22]

[14] ibid 10. [15] Palfrey and Gasser (n 9) 179. [16] ibid. [17] ibid 182.
[18] ibid 183. [19] ibid. [20] ibid 184. [21] ibid. [22] ibid 187.

9. 'States sometimes do not wish their legal regimes to interoperate with those of other states.'[23]
10. 'The choices we make with respect to legal interoperability can have public security consequences in a highly interconnected global economy.'[24]
11. 'The cost of noninteroperable laws in the highly networked world of global information and communication technologies will increase.'[25] and
12. 'Low levels of legal interoperability are likely to have negative long-term effects on innovation, trade, and economic growth.'[26]

All this applies with equal force to what I discuss as jurisdictional interoperability.

More specifically, if we want to make serious progress with jurisdictional interoperability, I suggest that we need to start by adopting a common understanding of—a common framework for—how we approach jurisdiction. And I suspect that no reader will be surprised if I here reconnect to the framework I advanced earlier. Adopting that framework, or something like it, will go a long way towards achieving jurisdictional interoperability.

At any rate, if we were to provide a definition of jurisdictional interoperability, we may say that it is achieved where we have: (1) only a minimal level of serious jurisdictional clashes, and (2) an acceptable level of less serious jurisdictional clashes. For these purposes, 'jurisdictional clashes' include both clashes of duties, and clashes between rights and duties. In my view, it is this goal we currently should work towards in the context of Internet jurisdiction. In doing so, we will, unsurprisingly, have a better prospect of success in some areas than in others. As a result, we may well end up with what may be referred to as 'sectorial jurisdictional interoperability' in some areas of law.

5.3 Concluding Remarks Regarding Jurisdictional Interoperability

This chapter draws on the conclusions reached in the previous chapter on the history of Internet jurisdiction and suggests that the most appropriate path forward at this stage is to focus our attention on the concept of 'jurisdictional interoperability' as a part of 'legal interoperability'. In other words, as already noted, rather than vainly hoping for an all-encompassing international agreement overcoming the problems of Internet jurisdiction, we must accept instead that the road ahead will be travelled by many thousands of

[23] ibid. [24] ibid 189. [25] ibid 191. [26] ibid.

small steps. Just as the Internet is a successful network of networks, the mid-term solution to the jurisdictional issues online will be found in what we can see as a *system of legal systems*—a system in which our domestic legal systems operate smoothly together with a minimum of inconsistencies and clashes.

Importantly, this means that rather than sitting back waiting for a miraculous international agreement addressing all the jurisdictional concerns online, everyone can get involved—law reformers, courts, legislators, lawyers, legal academics, civil society, and law students—in identifying uniting features[27] and in chipping away at the inconsistencies, contradictions, and clashes that hinder interoperability between the various legal systems that govern our conduct online. Getting such work underway is a matter of urgency, and it may well be the case that aspects of cosmopolitanism may prove to be a useful source of inspiration, or even guidance. That is, however, a topic I will not pursue further here.

[27] In this context, we would do well to recognise that even in controversial areas, such as data privacy law, unifying ideas can also be found. See further Lee A Bygrave, *Data Privacy Law: An International Perspective* (OUP 2014). Thus, while the road ahead will not be easy, one must not lose hope of interoperability.

6

Understanding the Functions of Jurisdictional Law

To be in a position to consider how we may shape our jurisdictional rules better to cater for jurisdictional interoperability, we need first to understand the functions those rules fulfil. And here we need to enter a hostile environment: the treacherous quagmires that are referred to as 'legal theory' and 'legal philosophy'. After all, to understand the functions of jurisdictional rules we need to understand the function of the law as such. This is, however, a huge topic that has occupied some of the sharpest legal minds in history. Consequently, the reader should not expect what follows to paint any full picture of the debates the world has witnessed in that arena. Rather, I have had to be extremely selective in what I have focused on, and I will not let this discussion detain us for long.

Elsewhere,[1] I have elaborated on what are the *aims* of private international law rules in general. Most of those aims may well be said to be the aims of jurisdictional law more broadly. Thus, as I see it, we can take account of at least the following when discussing what are the aims of jurisdictional law:

1. *Legitimate party expectations*—it is of fundamental importance that good jurisdictional rules be equipped to meet both genuine party expectations and constructive party expectations. Therefore, such rules must contain a context-dependent balance between predictability and flexibility.[2]
2. *Suitable level of generalisation*—it is of fundamental importance that good jurisdictional rules, just as any other rules, be drafted at a suitable level of generalisation.[3]

[1] Dan Svantesson, *Private International Law and the Internet* (3rd edn, Kluwer Law International 2016) 94–141.
[2] ibid 101–14. [3] ibid 114–15.

3. *Abuse-resistance*—it is of fundamental importance that good jurisdictional rules, like any other rules, are drafted in a manner that minimises the risk of abuse of those rules.[4]
4. *Non-violation of public international law*—it is of fundamental importance that good private international law rules do not go beyond what is allowed under public international law.[5]
5. *Non-violation of human rights law*—it is of fundamental importance that good jurisdictional rules do not go beyond what is allowed under human rights law.[6]
6. *Policy fulfilment*—like other rules, the jurisdictional rules must work towards the fulfilment of relevant policies.[7]
7. *Forum-neutral language*—to be applied in a consistent manner, jurisdictional rules likely to be applied by the courts of different jurisdictions must be drafted in a forum-neutral manner.[8]
8. *Efficiency*—like other rules, jurisdictional rules must be as efficient as possible.[9]
9. *Simplicity*—while difficult to achieve, jurisdictional rules must be as simple as possible.[10]
10. *Wide acceptance*—as with other legal rules, jurisdictional rules benefit from being widely acceptable among the stakeholders they affect.[11]
11. *Independence from substantive law concepts*—it is desirable that jurisdictional rules are not anchored in concepts and terminology stemming from substantive law.[12]
12. *Sufficient specificity*—it is desirable that jurisdictional rules are drafted with sufficient specificity to be applied in a consistent manner.[13]

However, these qualities are not all of equal importance, and it is useful to distinguish between fundamental and non-fundamental qualities.[14] The fundamental qualities are such that a good jurisdictional rule cannot operate without them. Put differently, the fundamental qualities represent the *limits to any acceptable solution*. The non-fundamental qualities are those that are desirable for every good jurisdictional rule but not essential. Of these, the

[4] ibid 115–17. [5] ibid 117–30. [6] ibid 130–32. [7] ibid 132–36.
[8] ibid 136–37. [9] ibid 137–38. [10] ibid 138. [11] ibid 139.
[12] ibid 139–40. [13] ibid 140–41.
[14] I have discussed the theoretical thinking behind the manner in which such qualities are to be identified elsewhere. See Dan Svantesson, 'A Legal Method for Solving Issues of Internet Regulation; Applied to the Regulation of Cross-Border Privacy Issues' (European University Institute Working Paper LAW2010/18).

first five should be seen as fundamental while the others, but one, are always non-fundamental qualities. The seventh quality—forum-neutral language—is the odd one out in that it is fundamental in some contexts and not in others. Further, the fourth quality is of course only of relevance in relation to private international law rules.

Before proceeding, I will pause to consider in more detail the fifth quality, that is, the non-violation of human rights law. In some recent discussions there has been a tendency to misconceive the relationship between human rights law and jurisdictional law. As noted, good jurisdictional law cannot violate human rights law and ought in fact to promote human rights law. However, it is a mistake to use jurisdictional rules as a substitute for substantive human rights law. Doing so is akin to use a height restriction as a substitute for an age restriction at the entrance to, for example, a pub. Setting an appropriate height restriction may keep out most people under the age of 18, but there will be both false positives and false negatives. Similarly, using jurisdictional restrictions to protect substantive human rights will probably result in both leakage and sub-optimal jurisdictional rules.

Be that as it may, I think there are reasons to dig a bit deeper here on the functions of jurisdictional law.

6.1 What Is Law?

A useful, albeit somewhat daunting, starting point here is to consider what law is. The answer will inevitably affect how we view our jurisdictional rules and their functions. After all, there can be no doubt that some types of jurisdictional rule may be more palatable in certain philosophical arenas than in others.

I argue that law *L*—as the subject-matter of legal theory—may be expressed as $n(Norm + {}_{Norm}(Context))$, where *Norm* represents legal norms created by legislative enactments or court judgments, and ${}_{Norm}(Context)$ represents those considerations, internal to the legal norm, that affect the application of the legal norm, that is, the legal norm's context.[15] This 'context' is broad indeed and includes matters such as the linguistic/terminology context, the historical context, the political context, the philosophical context, the systematic context, the economic context, the

[15] This is a revised version of a proposal I first advanced in 2013 in Dan Svantesson, 'What is "Law", if "the Law" is Not Something That "Is"? A Modest Contribution to a Major Question' (September 2013) 26(3) Ratio Juris 456–85. I am grateful to von der Pfordten for his valuable feedback that has allowed me to improve my proposal.

announcement context, the application context, and the structural context.[16] Finally, *n* as used here represents the unascertainable number of legal norms that make up the law.

Viewing law in this manner allows us to make several observations of importance concerning how we approach jurisdictional rules. First, doing so shows that a strict separation between law and politics is an illusion. Furthermore, my theory accounts for the significant role played by general principles of law and morality; they are both part of the context and thus part of law. In addition, as will be discussed in greater detail, this theory of what is law highlights the crucial role played by interpretation of legal norms. In addition, I argue that, as public international law restrictions form part of the ${}_{Norm}(Context)$ of private international law norms, applying this theory of law highlights yet another reason why we cannot strictly separate private and public international law in the context of jurisdiction. Finally, a natural consequence of the theory of law I advance is that the law is 'underdeterminate'.[17] This is perhaps a particularly serious concern in the context of laws being applied to online conduct. After all, in no other context are we typically exposed to so many different laws, and the application of law to technology is riddled with examples of difficult choices having to be made by the courts when they apply the law to the novel Internet environment—I suggest that in no other area of law is it as common that the law's operation is uncertain as in relation to technology law. This further undermines the idea that we can tailor our conduct to comply with applicable laws, which in turn undermines many 'extraterritorial' claims of jurisdiction over Internet conduct.

6.2 Law, Facts, and Interest Balancing—the 'Tools of the Trade'

Looking then at how courts operate, and more specifically their 'tools of the trade', we may be able to link this discussion to the jurisprudential framework for jurisdiction advanced in the previous section. Indeed, I argue

[16] In fact, I argue that ${}_{Norm}(Context)$ takes account of both the past and the present, and also the future. Future decisions are part of the ${}_{Norm}(Context)$ now because they affect disputes arising now, even if the courts decide the matter only in the future (see further ibid 469). However, I have not failed to realise just how controversial this suggestion is, and to avoid it causing distraction, I do not raise it here. I will, however, make the very brief observation that I find it odd that, while physicists are engaged in serious discussions about parallel universes, legal scientists are so often refusing to entertain any ideas that challenge fundamental beliefs about the legal system.

[17] See ibid 456.

that the view of the law outlined here potentially provides support for that framework.

A court works with—and indeed, all legal decisions are based on—three 'tools'; *law, facts,* and *interest balancing.* These are the things—the only things—a properly functioning court considers in any given dispute, and these focal points survive the 'always and only test' in that they are always applied by every court and only they hold that status. That is, each and every legal question can be tackled by reference to these three focal points—law, facts, and interest balancing.

Elsewhere, Polák and I have suggested that, philosophically, the two categories—the 'substantial connection' requirement and the 'legitimate interest' requirement—are independent because the latter is normative and the former is factual.[18] If this is so, and it is conceded that the tools courts work with are indeed law, facts, and interest balancing, it may reasonably be suggested that the reason the three-principle jurisprudential framework for jurisdiction fits multiple contexts is found in the fact that those three principles are rooted in the three focal points for all legal inquiries.

Be that as it may, the observation that every legal question is tackled by reference to the three focal points of law, facts, and interest balancing allows us to reach at least one additional important conclusion. While there is a great appetite for packaging and promoting jurisdictional considerations as 'single-factor tests', such as placing focus on the location of contract formation or of the server, we may be deceiving ourselves as to the simplicity such tests actually provide. In this context, I make three claims of relevance for both the discussion to follow and the discussion already had about my proposed multifactor framework.

First, I argue that genuine single-factor tests are rarely, if ever, capable of reflecting the complexity of considerations involved in a jurisdictional dispute. Thus, they always represent a normative dumbing-down of the matter, and this may be the reason that many that appear to be single-factor tests are in fact nothing but badly disguised multifactor tests. Consider, for example, jurisdiction based on the centre of interests. This may, at a first glance, appear to be a single-factor test, but we need only scratch the surface and consider how we determine what amounts to a person's centre of interest to realise that we are here dealing with an obvious multifactor test. Or perhaps even more clearly, while the 'targeting' test may appear to be single-factor, behind its façade lies a complex multifactor test taking account of matters such as the currencies in which payments are accepted, the top-level domain used, and

[18] Dan Svantesson and Radim Polák, *Information Sovereignty* (Edward Elgar, forthcoming 2017). This arguably also connects to Hume's distinction between the 'is' and the 'ought'.

the language in which the website is presented.[19] These are consequently not genuine single-factor tests; they merely hide the multi-factor nature behind the veil of a single-factor façade.

Second, I suspect that where we opt for a genuine single-factor test, we lock ourselves into a certain position as far as interest balancing is concerned. In other words, through the decision to favour one single-factor test over another, we consciously—or more perilously, unconsciously—adopt a certain interest balance that we later are unable to modify within the single-factor test itself. For example, if we decide that the question of whether law enforcement may access data stored in a cloud arrangement on a server in another state is determined by reference to the nationality of the offender, we implicitly make certain value choices as to the interests to be balanced favouring some over others. And once we have locked ourselves into such a single-factor test, we can only modify that interest balance by means external to the chosen single-factor test.

Third, because of the loss of flexibility that comes with the selection of a genuine single-factor test, the overall enquiry that the test fits within may need to compensate for the lost opportunity to engage in a proper interest balancing within the single-factor test. Consider, for example, a genuine single-factor test such as jurisdiction based on the location of contract formation. Identifying where the contract is formed is a mechanical task guided by predetermined interest balancing choices that do not lend themselves to adjustment in the individual case. However, those who praise the simplicity and predictability that stem from such a mechanical single-factor test seem to overlook the fact that the application of that test is merely a component of a bigger picture. For example, even where a court finds that it may claim jurisdiction based on a contract having been formed within the State where the court sits, it may nevertheless be in a position to use the doctrine of *forum non conveniens* or the doctrine of *lis alibi pendens* to decline to hear the matter. Thus, we can conclude something very important—even where we opt for genuine single-factor tests that lend themselves to truly mechanical application, that alone does not guarantee that the overall decision is simple and predictable since we cannot avoid taking account of law, facts, and interest balancing in properly addressing a legal matter. And this is precisely why both public and private international law are riddled with normative tools focused on interest balancing, such as 'ordre public', *forum non conveniens*, and comity.

[19] Joined Cases C 585/08 and C 144/09, at paras 83 and 93, as well as Opinion of Advocate General Trstenjak delivered on 18 May 2010, at para 84.

6.3 The Triple Role of Law

Much, indeed too much, has been made of the impact enforcement difficulties have on jurisdictional claims. For example, a leading commentator, Goldsmith, has stated:

Most Internet content providers will not be subject to any regulation other than the one in the territory in which they have presence. […] [T]hese Internet users might indirectly suffer consequences from another nation's territorial regulation of the user's Internet transmission. But these offshore users with no local assets are generally beyond the regulating nation's enforcement jurisdiction. The Internet users that need to worry about the liability consequences of multiple, conflicting regulatory requirements are persons and firms with a multi-jurisdictional presence.[20]

Goldsmith uses this reasoning to conclude that threats of simultaneous multiple national regulation of Internet transactions are significantly exaggerated. I do not fully share Goldsmith's assessment.[21] The real litmus test for the value of jurisdictional claims is not to be found in whether such claims can be backed up by enforcement. It is not entirely correct as Goldsmith states that '[t]he true scope and power of a nation's regulation is measured by its enforcement jurisdiction, not its prescriptive jurisdiction.'[22] We may here recall what I see as one of Hart's most important contributions:

The principal functions of the law as a means of social control are not to be seen in private litigation or prosecutions, which represent vital but still ancillary provisions for the failures of the system. It is to be seen in the diverse ways in which the law is used to control, to guide, and to plan life out of court.[23]

This observation may seem mundane indeed, and Hart is perhaps not the only, or even the first, commentator to bring attention to what we can call 'the dual role of law', but it is one of the most important observations made in Hart's *The Concept of Law* and something we must always keep in mind.[24]

[20] Jack Goldsmith, 'Unilateral Regulation of the Internet: A Modest Defence' (2000) 11 Eur J Intl L 135–48, 139–40.

[21] However, I hasten to add that to evaluate properly what he is stating we must take note of two things: first, Goldsmith is conscious of the effect of what I term below 'market destroying measures', and second, he takes a traditional narrow view on what is extraterritorial and what is territorial with spillover.

[22] Goldsmith (n 20) 135, 139.

[23] HLA Hart, *The Concept of Law* (3rd edn, OUP 2012) 40.

[24] Similar notions may also be found in more recent commentaries. For example, MacCormick notes: 'Normative power is, then, the ability to take decisions that change what a person ought to or ought not to do, or may or may not do, or what a person is able or unable to do, in the framework of some normative order.' (Neil MacCormick, *Institutions of Law* (OUP 2007) 154.) The

In the light of the dual role of law, enforcement is not quite as essential a feature of law as one may first think on a superficial consideration of the matter. After all, the law's role as a tool to control, to guide, and to plan life out of court does not necessarily depend on enforcement. This is perhaps particularly so with international law, and we may draw upon one of the four answers Koskenniemi identifies as relevant for the question of 'what is international law for?':

> [It] is international law's formalism that brings political antagonists together as they invoke contrasting understandings of its rules and institutions. In the absence of agreement over, or knowledge of the 'true' objectives of political community—that is to say, in an agnostic world—the pure form of international law provides the shared surface—the *only* such surface—on which political adversaries recognise each other as such and pursue their adversity in terms of something shared, instead of seeking to attain full exclusion—'outlawry'—of the other. In this sense, international law's value and its misery lie in its being the fragile surface of political community among states, other communities, individuals who disagree about their preferences but do this within a structure that invites them to argue in terms of an assumed universality.[25]

Thus, as an expanded aspect of Hart's assertion of law's role as a tool to control, to guide, and to plan life out of court, we may add the function of being the surface—playing board—on which political adversaries recognise each other as such and pursue their adversity in terms of something shared.

Either way, putting all this in the context of claims of jurisdiction then, the real litmus test, I argue, is the extent to which a State's jurisdictional claim can affect the conduct of foreign parties—whether the consequences of a breach of State A's laws is seen, by the targeted foreign actors, as having consequences of such a nature as to make those actors prefer abiding by State A's law, rather than having to face those consequences. In this context, the potential enforcement is only one of the relevant factors (another being, for example, the potential for reputational damage in case of breach).

The potential for reputational damages is of course quite self-explanatory and has been neatly described by Kohl in these terms: 'The fact is that being perceived as a law-breaker is not good for business.'[26] To this I would,

reference to normative power as the ability to take decisions that change what a person ought or ought not to do clearly relates to law as a tool to guide and to plan life out of court. In contrast, the reference to normative power as the ability to take decisions that change what a person is able or unable to do is more in line with law's other role.

[25] Martti Koskenniemi, *The Politics of International Law* (Hart 2011) 266.

[26] U Kohl, *Jurisdiction and the Internet: Regulatory Competence over Online Activity* (CUP 2007) 208.

however, add the important caveat that the damage done by being perceived as a lawbreaker depends on whether the law being violated is seen as morally justifiable or not. For example, the international reputation of a company that has been breaching the racial discrimination rules of Nazi Germany and of apartheid South Africa may indeed be boosted by the lawbreaking activity.

In any case, in Hart's dual role of law can we already find support for my claim that Goldsmith and his disciples overrate the significance of effective enforcement at the expense of providing a more holistic picture of what law aims to achieve. To this we may add what I see as a third role that the law serves. In my view, we can say that quite apart from being a tool to decide legal disputes and to provide a framework to control, to guide, and to plan life out of court, law is a tool to communicate the values of the society that created the law.[27] This is important not least in the context of the discussion of what I below refer to as 'bark jurisdiction'.

Where all this is conceded, we may conclude that law fulfils at least three different roles. Law is: (1) a tool to decide legal disputes, (2) a tool to provide a framework to control, to guide, and to plan life out of court, and (3) a tool to express and communicate the values of the society that created the law.[28] Importantly, only the first of these roles is directly dependent on physical enforcement and therefore on territoriality. This—what I call the 'triple role of law'—again emphasises that claims that territoriality necessarily imposes limitations on prescriptive jurisdiction due to the dependence on territorial enforcement jurisdiction are misguided.[29]

Arguably, the recognition of the triple role of law is of particular significance in the context of international law. In his dissenting opinion in *Lotus*, Loder spoke of international law in these terms:

> This law is for the most part unwritten and lacks sanctions; it rests on a general consensus of opinion; on the acceptance by civilized States, members of the great community, of nations, of rules, customs and existing conditions which they are bound to respect in their mutual relations, although neither committed to writing nor confirmed by conventions. This body of rules is called international law.[30]

[27] Dan Svantesson, 'A Jurisprudential Justification for Extraterritoriality in (Private) International Law' (2015) 13(2) Santa Clara J Intl L 517–71, 551–52.

[28] Two of these roles may be derived from Hart (n 23) 40: 'The principal functions of the law as a means of social control are not to be seen in private litigation or prosecutions, which represent vital but still ancillary provisions for the failures of the system. It is to be seen in the diverse ways in which the law is used to control, to guide, and to plan life out of court.'

[29] See eg Cedric Ryngaert, *Jurisdiction in International Law* (2nd edn, OUP 2015) 195, arguing that 'prescriptive jurisdiction can only be effective if it is matched with enforcement jurisdiction—which is under the current international rules only permissible if *territorially* exercised' (emphasis added).

[30] SS 'Lotus' (*France v Turkey*) (1927) PCIJ Series A, No 10, 34.

Indeed, the fact that a lack of enforcement mechanisms is somewhat of a hallmark of international law is widely acknowledged. As noted by Guzman, 'Whatever the strengths of international law, it remains almost entirely without coercive enforcement—the primary tool used to generate compliance in domestic systems.'[31] This arguably sets this legal arena apart from some truths otherwise generally applicable to the legal system as such. Here, it is also worth noting how, in discussing the criticism raised against Kelsen's perception of public international law, Somek has stated: '[i]t could be argued that even if all states disrespected their international obligations these obligations would be *effective enough* for the international legal system to exist as long as universal non-compliance is universally *perceived* as legally wrong. The whole system would be, at best, a system of universal hypocrisy.'[32] This statement is no less relevant for our context here. In the end, the true value, and certainly legitimacy, of jurisdictional claims may not rest on the enforcement, or indeed enforceability of the claim. Instead, in our system of 'universal hypocrisy', such claims have both value and jurisprudential legitimacy as long as non-compliance is *perceived* as legally wrongful. And I would argue that such a perception need not necessarily be universal in nature, as long as it is sufficiently widespread.

6.4 Bite versus Bark Jurisdiction

The findings of the previous section have important implications for how we view the value of jurisdictional claims. It is standard practice in literature dealing with jurisdictional matters to emphasise the harmful effect we may expect from jurisdictional claims that cannot be enforced. Most, if not all, scholars of note on the Internet jurisdiction arena appear to subscribe to this notion.

Bygrave, for example, has discussed the issue using the term 'regulatory overreaching': 'By "regulatory overreaching" is meant a situation in which rules are expressed so generally and non-discriminately that they apply *prima facie* to a large range of activities without having much of a realistic chance of being enforced.'[33] Like several other leading commentators, such as Kuner,[34]

[31] Andrew T Guzman, *How International Law Works: A Rational Choice Theory* (OUP 2008) ix.

[32] Alexander Somek, 'Kelsen Lives' (2007) 18 Eur J Intl L 409, 430 (internal footnote omitted).

[33] Lee Bygrave, 'Determining Applicable Law Pursuant to European Data Protection Legislation' (2000) 16 Compu L & Sec Rep 252, 255.

[34] Christopher Kuner, 'Data Protection Law and International Jurisdiction on the Internet (Part 2)' (2010) 18(3) Intl JL & Info Tech 227, 235.

Trimble,[35] Maier,[36] Reed,[37] and Moerel,[38] Bygrave sees 'regulatory overreaching' as a problem,[39] although perhaps one can see hints of a softening in Bygrave's approach in his more recent writings. Writing about extraterritoriality in data privacy law, Bygrave states that he is 'sceptical of giving law an extraterritorial dimension that remains dormant.' But adds that '[i]t could be argued, however, that giving law an extraterritorial reach that is not followed up by practical enforcement is still valuable as a demarcation of jurisdictional lines'[40]

Given what I have said above regarding the triple role of law, I regret that I am unable to concur with the opinion of these scholars whose views I hold in high esteem. Elsewhere,[41] I have introduced the distinction between what I call 'bite jurisdiction', on the one hand, and 'bark jurisdiction' on the other.

The way I see the matter, not all jurisdictional claims are equally likely to be carried out in practice. Indeed, some jurisdictional claims are made

[35] M Trimble, 'Advancing National Intellectual Property Policies in a Transnational Context' (2014–15) 74 Md LR 203, 234.

[36] Bernhard Maier, 'How Has the Law Attempted to Tackle the Borderless Nature of the Internet?' (2010) 18 Intl JL & Info Tech 161.

[37] Chris Reed, *Making Laws for Cyberspace* (OUP 2012) 49: 'The enforcement of a law also plays an important role in engendering respect for that law. If a state consistently fails to enforce a law it sends a message to the law's subjects that the state does not expect them to obey it. [. . .] Where a state wishes to enforce a law but is unable to do so in any consistent and effective way, the message is rather different but equally damaging to respect. Here the state desires that its commands should be obeyed, but is impotent to force individuals to do so. This weakens respect not merely for the particular law but for *all* that state's laws.' Further, in another publication, Reed stresses: 'A regulator which is otherwise accepted as having legitimate authority can easily lose that authority if it has no effective way of enforcing its rules' (Chris Reed, 'Cloud Governance: The Way Forward' in Christopher Millard (ed), *Privacy Law: Comparative Perspectives* (OUP 2013) 374. Elsewhere in his interesting book, *Making Laws for Cyberspace*, Reed seeks to incorporate enforceability into a 'rule of recognition' for Cyberspace, like the rule of recognition famously discussed by Hart: 'Adopting Raz's approach would suggest that what we might term the subject rule of recognition has two limbs: foreign laws have authority for a cyberspace actor if the risk [. . .] of their enforcement (or other adverse consequences) is sufficiently great that a prudent actor would follow them; and they also have authority if the actor has sufficient respect for that foreign legal system to create a moral obligation to comply with its laws, or at least some of them' (Reed (n 37) 86). Combining these quotes it seems that, while Reed has clear concerns about bark jurisdiction causing a loss of respect for the law, he nevertheless admits that bark jurisdiction may well meet the criteria of his rule of recognition.

[38] Lokke Moerel, 'The Long Arm of EU Data Protection Law: Does the Data Protection Directive Apply to Processing of Personal Data of EU Citizens by Websites Worldwide?' (2011) 1 Intl Data Priv L 23, 24.

[39] Bygrave (n 33) 255.

[40] Lee Bygrave, 'Data Privacy Law and the Internet: Policy Challenges' in Normann Witzleb, David Lindsay, Moira Paterson, and Sharon Rodrick (eds), *Emerging Challenges in Privacy Law: Comparative Perspectives* (CUP 2014) 277, referring to the bite/bark distinction I outlined in Dan Svantesson, *Extraterritoriality of Data Privacy Law* (Ex Tuto 2013) 83–88.

[41] Svantesson (n 40) 83–88.

despite the realisation that they have virtually no prospect of being exercised in practice—they are merely attempts to 'bark': to make clear—to articulate—a particular legal position. Placed in Hart's terminology, they fall in the category of law used to control, to guide and to plan life out of court. Or, indeed, as I have argued, they may be made to communicate the values of the society that made the claims. In contrast, jurisdictional claims properly described as 'bite jurisdiction' are aimed at actually being effectively enforced. Obviously, it may not always be easy to draw sharp lines between failed attempts at bite jurisdiction, on the one hand, and genuine instances of bark jurisdiction on the other.

The difficulty of distinguishing between failed attempts at bite jurisdiction from genuine bark jurisdiction brings us to the first matter that must be discussed in some depth—do, and should, states make clear whether the jurisdictional claims they make are bite or bark?

My suspicion is that, currently, there is little, indeed too little, consideration amongst those making jurisdictional claims, as to whether they are making a bite or bark claim. However, exceptions can be found. For example, in discussing the extraterritorial dimension of the then proposed Singaporean Personal Data Protection Bill (PDPB), the Ministry of Information, Communications and the Arts (MICA) observed:

> MICA is cognisant of the implementation challenges. In particular, where the organisation in question has no presence in Singapore, it would be difficult to carry out investigations into any complaint made in relation to an activity of the organisation, or to proceed with any enforcement action against the organisation. However, such coverage would act as deterrence for overseas companies to engage in activities that might result in a breach of the PDPA, and provide consistent treatment for local vis-a-vis overseas organisations with data-related operations in Singapore.[42]

It would thus seem that there is an expectation that the anticipated grab for 'extraterritoriality', in part, is a conscious, and express, claim for 'bark jurisdiction' rather than a real attempt at 'bite jurisdiction'. Furthermore, while the Australian consumer law has some extraterritorial reach, the body tasked with enforcing it—the Australian Competition and Consumer Commission (ACCC)—seems to view this largely as an instance of a mere bark claim: 'All your usual consumer rights apply when you shop with an Australian online business. Those rights may also apply when you buy from an overseas online business although you might find it difficult to get a repair, replacement

[42] *Public Consultation Issued by Ministry of Information, Communications and the Arts—Proposed Personal Data Protection Bill* (19 March 2012).

or refund because the business is not based in Australia.'[43] Thus, it is clear that in some instances bark claims are, more or less, clearly identified as such, and perhaps the bark nature is quite widely accepted amongst the public. Nevertheless, any decision as to whether to announce an extraterritorial claim as bark jurisdiction must be guided by a balancing of the advantage of transparency and the risk that such an announcement will cause the claim to be ignored.

In the light of all this, I do not see it as accurate to view 'bark jurisdiction' or 'regulatory overreaching' as a problem *per se*. After all, there may well be solid reasons why a state may wish to make clear its standpoint on a particular issue by legislating against it, even though the effective enforcement of the law in question may be difficult, cumbersome or, indeed, unlikely. To see that this is so, we need only consider laws against domestic violence. Statistics prove that such laws are not always deterring offenders, and more to the point, that such laws are not effectively enforced. But would anyone seriously suggest that this is a reasonable basis to conclude that we therefore do not need such laws, or indeed, that they, due to their limited enforcement, undermine the law? I hope not, and it seems beyond intelligent dispute that the value of these laws is partly found in the societal values they communicate.

This is as true in the international jurisdictional context as it is in the context of domestic law with a clear territorial limitation (eg legislation making it a criminal offence to drive against a red light). Turning to the data privacy law environment, it is worth pausing to consider how the highly influential EU Article 29 Working Group has noted: '[T]here exist examples that the foreign web site may nevertheless follow the judgement [sic] and adapt its data processing with a view to developing good business practice and to maintaining a good commercial image [even where a third countries will not recognise and enforce the judgment.]'[44] We can here also reconnect to the reputational dimension brought into focus earlier. Support for this proposition can also be drawn from Hart's statement: 'In fact I think it quite vain to seek any more specific purpose which law as such serves beyond providing guides to human conduct and *standards of criticism of such conduct*.'[45] Hart is here, in the italicised text, seemingly referring to the law's role as 'bark'

[43] Australian Competition and Consumer Commission's website <http://accc.gov.au/consumers/online-shopping/shopping-online>, accessed 4 April 2017.

[44] Article 29 Working Group, 'Working Document on Determining the International Application of EU Data Protection Law to Personal Data Processing on the Internet by Non-EU Based Web Sites' (Working Paper 56, adopted on 30 May 2002) 15.

[45] Hart, *The Concept of Law*, 249 (emphasis added).

jurisdiction; bark jurisdiction is all about providing standards of criticism of particular kinds of conduct.[46]

Thus, I see the point Kohl makes when she claims that 'large-scale non-compliance with any legal rule is problematic not just in terms of the failure of achieving the law's purpose, but also in terms of undermining the law's and regulator's credibility more generally'[47] and Kuner's similar claim that laws that lack the means of being enforced can be seen to undermine the legal system.[48] And I can sympathise to a degree with Bygrave's assertion that 'posturing without punch or even potential punch tends to be counterproductive',[49] and statements such as that '[w]hen criminal laws have nothing but symbolic value [. . .] they are likely to erode rather than build confidence in the justice system, since they quickly come to be seen as paper tigers.'[50] However, I am not convinced these claims go deep enough to do justice to the complexity of the issue at hand. On my reading, they seem to (1) neglect or underestimate[51] the triple role of law and (2) undervalue the reputational dimension of legal compliance. And indeed, in using the example of how also ineffective regulation of domestic violence has legitimacy and utility, I hope that I have brought attention to what must be a fundamental flaw of statements such as those noted earlier.

Adding to what I have already remarked on this issue, I suggest that the risk that laws that lack the means of being enforced will undermine the legal system is small where the parts of the law that are difficult to enforce are not dominant or even close to being the dominant feature of the legal system in question. One need only consider those situations where the population in abusive dictatorships cling onto the notion of human rights even though those rights are unlikely to be upheld; morally justifiable law[52]—including morally justifiable law that cannot be enforced—has a quality that cannot,

[46] This point is in no way undermined by the fact that Hart follows the quoted statement with the observation that: '[t]his will not of course serve to distinguish laws from other rules or principles with the same general aims; the distinctive features of law are the provision it makes by secondary rules for the identification, change, and enforcement of its standards and the general claim it makes to priority over other standards' (ibid).

[47] Kohl (n 26) 153.

[48] Kuner (n 34) 235–36, making reference to works such as Hart, *The Concept of Law* and Hans Kelsen, *General Theory of Law and State* (Transaction 2005).

[49] Bygrave (n 40) 277.

[50] S Coughlan et al, 'Global Reach, Local Grasp: Constructing Extraterritorial Jurisdiction in the Age of Globalization' (2007) 6 Can JL & Tech 32, 50.

[51] Bygrave (n 40) 277 acknowledges the dual role of the law by observing that '[d]emarcation of values and ideals is an integral element of all law.' However, he then returns to his starting point of a need for the law to 'pack a punch'.

[52] Defining this term would take us into an interesting and fascinating area of debate that unfortunately goes beyond the scope of this chapter.

and should not, be ignored. And maybe this is exactly where we reach the core of this issue—moral justification.

6.4.1 When is Bark Jurisdiction Appropriate?

In her 2007 book on jurisdiction and the Internet, Kohl states: 'It is enforceability that really matters, not actual enforcement.'[53] She then proceeds to note that, at least in the transnational context, the reason for the importance of enforceability 'lies often not simply, or even mainly, in inducing a fear of a sanction in the case of non-compliance, but rather in affirming the foreign law's legitimacy'.[54]

Thus, perhaps it can be said that the relevance and value of bark jurisdiction depends on whether the jurisdictional claim, and the substantive law it relates to, is morally justifiable.

Expanding on the reputational dimension introduced earlier, we can say that where the bark jurisdiction and the substantive law it relates to is morally justifiable, it is perilous for the target of the claim to ignore it; and where the bark jurisdiction and/or the substantive law it relates to is not morally justifiable, it is perilous for the country making the claim to make the jurisdictional claim. Revealing my idealistic (or perhaps naive) side, it may perhaps be said that this should have the dual positive effect of encouraging restraint amongst countries considering making too broad extraterritorial claims and should encourage compliance with rules that otherwise may have been ignored amongst the targets of the extraterritorial claims. But in Bygrave's later writings we find a statement that should remind us that the landscape before us is more complicated than I so far have let on.

Having noted how '[d]emarcation of values and ideals is an integral element of all law,'[55] Bygrave states that 'posturing without punch' may be particularly counterproductive to the general respect for those values where 'the values are not otherwise widely respected *in practice*'.[56] In the foregoing, I have linked the moral dimension of the extraterritorial claim with the morality of the substantive law. This has given us two options, as illustrated in Table 6.1.[57]

The reality, however, is of course better described in Table 6.2.[58]

This somewhat increased sophistication forces us to consider the matter before us with greater precision. How do we view bark jurisdiction in these

[53] Kohl (n 26) 205. [54] ibid. [55] Bygrave (n 40) 277. [56] ibid.

[57] Dan Svantesson, 'A Jurisprudential Justification for Extraterritoriality in (Private) International Law' (2015) 13(2) Santa Clara J Intl L 517–71, 562.

[58] ibid.

Table 6.1 Morality in extraterritoriality

(x) Extraterritorial claim moral and substantive law moral.	(y) Extraterritorial claim immoral and/or substantive law immoral.

Table 6.2 Morality in extraterritoriality

(a) Extraterritorial claim moral and substantive law moral.	(b) Extraterritorial claim immoral and substantive law moral.
(c) Extraterritorial claim moral and substantive law immoral.	(d) Extraterritorial claim immoral and substantive law immoral.

Table 6.3 Morality in extraterritoriality

		Extraterritoriality		
Substantive law		Moral	Dubious	Immoral
	Moral	V	?	X
	Dubious	?	?	X
	Immoral	X	X	X

scenarios? Scenario *a* poses no problems—it remains the case that where both the extraterritorial claim, and the substantive law it is seeking to give application to, are morally justifiable, it is perilous for the target of the claim to ignore it. The scenarios labelled *b, c,* and *d* are, however, more complicated. Most interestingly, perhaps it could be said that an immoral extraterritorial claim, such as where there is a lacking nexus between the action and the claim, may undermine the substantive law it seeks to make applicable even where that substantive law is morally justifiable. I will happily concede that, in such a situation (scenario *b* in Table 6.2), bark jurisdiction is inappropriate; but then so is bite jurisdiction. The same can be said for scenarios *c* and *d*. It is only for situations that fit scenario *a* that I advocate bark jurisdiction.

But perhaps also the expanded matrix is too simplistic to do justice to the complicated matter before us. Perhaps we get closer to a satisfactory model if we expand it further, as in Table 6.3.[59]

This matrix acknowledges that it is too simplistic to view, for example, an extraterritorial claim as either moral or immoral; rather the grey zone may be quite dominant with only limited black and white areas.

[59] ibid 563.

6.4.2 Bark Jurisdiction and the Right to an Effective Remedy

One last problem must be encountered before we draw any conclusions from the discussion here—does bark jurisdiction fit with the need to provide individuals with an effective remedy? As noted by Kuner in private correspondence,[60] one of the fundamentals of, for example, European human rights law is that individuals must have a legal remedy for violations of the law (at least in terms of the fundamental rights provided under such human rights law). For example, according to Article 13 of the European Convention on Human Rights (ECHR): 'Everyone whose rights and freedoms as set forth in this Convention are violated shall have an effective remedy before a national authority notwithstanding that the violation has been committed by persons acting in an official capacity.'[61] If the lawmaker enacts legislation that it knows in advance cannot be enforced and thus provides no feasible legal remedy, does that not raise questions about whether it has complied with basic principles of legality?

This issue is of both the greatest significance and the highest complexity. And importantly, as Kuner also has brought to my attention, it is not merely an academic matter; it is a matter that has distinct practical implications.[62] Further, it is, of course, not merely a European problem. Article 3 of the International Covenant on Civil and Political Rights[63] (ICCPR) states:

Each State Party to the present Covenant undertakes:
(a) To ensure that any person whose rights or freedoms as herein recognized are violated shall have an effective remedy, notwithstanding that the violation has been committed by persons acting in an official capacity;

60 Email of 19 March 2014 (on file with author).

61 See eg Convention for the Protection of Human Rights and Fundamental Freedoms, Rome, 4 XI 1950, Art 13.

62 Email of 28 July 2014 (on file with author). To keep this detour as brief as possible, and to avoid introducing the type of unnecessary confusion that inevitably comes with half-hearted attempts at describing matters of great complexity, we may here note, only in the briefest and most superficial terms, the recent judgment by the CJEU in Joined Cases C-293/12 and C-594/12. There, the Court noted: 'In the second place, it should be added that that directive [Directive 2006/24/EC of the European Parliament and of the Council of 15 March 2006 on the retention of data generated or processed in connection with the provision of publicly available electronic communications services or of public communications networks and amending Directive 2002/58/EC] does not require the data in question to be retained within the European Union, with the result that it cannot be held that the control, explicitly required by Article 8(3) of the Charter, by an independent authority of compliance with the requirements of protection and security, as referred to in the two previous paragraphs, is fully ensured. Such a control, carried out on the basis of EU law, is an essential component of the protection of individuals with regard to the processing of personal data [...]' (Joined Cases C-293/12 and C-594/12, para 68).

63 International Covenant on Civil and Political Rights, GA Res. 2200A (XXI), 21 UN GAOR Supp No 16, UN Doc A/6316/1966, 999 UNTS 171 (entered into force 23 March 1976).

(b) To ensure that any person claiming such a remedy shall have his right thereto determined by competent judicial, administrative or legislative authorities, or by any other competent authority provided for by the legal system of the State, and to develop the possibilities of judicial remedy;
(c) To ensure that the competent authorities shall enforce such remedies when granted.

Furthermore, Article 2(1) of the ICCPR states:

> Each State Party to the present Covenant undertakes to respect and to ensure to all individuals within its territory and subject to its jurisdiction the rights recognized in the present Covenant, without distinction of any kind, such as race, colour, sex, language, religion, political or other opinion, national or social origin, property, birth or other status.

It seems possible to argue that the phrase 'to respect and to ensure to all individuals *within its territory and subject to its jurisdiction* the rights recognized in the present Covenant' expresses two separate requirements rather than a double requirement.[64] As I have argued elsewhere,[65] from that vantage point, each signatory State has an obligation to provide legal protection against unlawful attacks on the rights of people subject to its jurisdiction *and* those present within its territory, regardless of the origins of the attacks.

This interpretation is supported in ICCPR General Comment 16: 'Provision must also be made for everyone effectively to be able to protect himself against *any* unlawful attacks that do occur and to have *an effective remedy against those responsible*'.[66] If, for example, the reputation of a person in State B is negatively affected by material posted on a website in State A, State B is arguably failing to provide 'an effective remedy against those responsible' unless its laws provide for extraterritorial jurisdictional and legislative claims over the offender in State A. However, what is then the impact of difficulties in enforcing the extraterritorial claim? It can, of course, be said that even such a jurisdictional claim does not in itself provide 'an effective remedy against those responsible' unless it can also be enforced. However, it would seem counterintuitive if State B in our example was to be required by international law to do more than what lies in its power to do. Parallels may here be seen to Dworkin's reasoning in his posthumous article on the basis of international law:

[64] M Nowak, *UN Covenant on Civil and Political Rights* (NP Engel 1993).
[65] See eg Svantesson (n 40).
[66] UN Human Rights Committee (HRC), *CCPR General Comment No. 16: Article 17 (Right to Privacy), The Right to Respect of Privacy, Family, Home and Correspondence, and Protection of Honour and Reputation*, 8 April 1988 <http://www.refworld.org/docid/453883f922.html>, accessed 4 April 2017 (author emphasis).

If a state can help to facilitate an international order in a way that would improve the legitimacy of its own coercive government, then it has a political obligation to do what it can in that direction. Of course that obligation demands only what, in the circumstances, is feasible. It does not require any state to ignore the division of the world into distinct states and suppose that it has the same responsibilities to citizens of other nations as it has to its own.[67]

Perhaps the issue is a rather simple one after all; State B in our example has two alternatives. The first option is to make an extraterritorial claim knowing that it may have a mere bark effect. The second is to refrain from giving extraterritorial effect to its laws regulating defamation. If it adopts this latter option, it avoids the criticism of having made law it cannot back up with an effective remedy. However, it is of course no closer to complying with the requirements of the ECHR or the ICCPR—after all, it then does not even make an attempt to uphold the relevant fundamental human rights against attacks originating abroad as it is obligated to do. Surely the conclusion must be that, at the minimum, it cannot be more wrong to try but fail to provide effective remedy than it is not to even try. If anything, one would think that, in this setting, bark jurisdiction is a better option than is complete and unconditional surrender to extraterritorial attacks.

In the light of this, my conclusion is that jurisdictional claims that can be seen as bite-less bark do indeed serve a function. Perhaps it could be said that 'bark jurisdiction' signals a perceived right to regulate a particular matter while acknowledging the lack of ability to regulate that matter. Viewed from this perspective, I argue that bark jurisdiction has both jurisprudential legitimacy and practical utility. And while I would happily lead the charge in any attempt to subject jurisdictional claims to appropriate checks and balances—not least as it applies to the Internet—I do not think an insistence on effective enforceability amounts to an appropriate yardstick for jurisdictional claims.

6.5 Market Sovereignty Based on Market Destroying Measures

When the claim is made, as it often is, that the online environment is associated with enforcement difficulties, we are not always very precise about to what we refer. First of all, we have reason to distinguish between difficulties of identifying the offending party, on the one hand, and difficulties of

[67] Ronald Dworkin, 'A New Philosophy for International Law' (2013) 41 Phil & Pub Aff 2, 17.

enforcing the law against the offending party on the other. But in the context of the latter, we also need to distinguish between foreign enforcement of a claim of jurisdiction and, in contrast, domestic enforcement of a jurisdictional claim. The problems associated with the foreign enforcement of a claim of jurisdiction may stem from difficulties in bringing the foreign party before the court or, for example, difficulties associated with having a judgment recognised abroad.

I suspect we may have paid too much attention to such difficulties at the expense of properly recognising the great deal of domestic enforcement that may be taken to give efficiency to a jurisdictional claim. A statement by Goldsmith and Wu is illustrative (although taken somewhat out of context). They observed that '[w]ith few exceptions governments can use their coercive powers *only* within their borders and control offshore Internet communications *only* by controlling local intermediaries, local assets, and local persons.'[68]

I have no quarrels with the factual correctness of this statement. Nevertheless, we may perhaps better describe the true state of things, and do so with greater clarity, if we remove the word 'only' from the quoted statement made by Goldsmith and Wu so as to end up with the following sentence instead: 'with few exceptions, governments can use their coercive powers within their borders and control offshore Internet communications by controlling local intermediaries, local assets, and local persons'. This alters the statement from what can be called, somewhat harshly, a meaningless cliché to a highly useful description of principles well-established at least 400 years ago.[69]

The word 'only' misleadingly gives the impression that such powers are of limited, or no, significance for the overall question of jurisdiction. After all, the power governments have within their territorial borders can be put to great effect against offshore Internet communications. A government determined to have an impact on foreign Internet actors that are beyond its directly effective jurisdictional reach may introduce what we can call 'market destroying measures' to penalise the foreign party. For example, it may introduce substantive law allowing its courts, due to the foreign party's actions and subsequent refusal to appear before the court, to make a finding that:

- that party is not allowed to trade within the jurisdiction in question;
- debts owed to that party are unenforceable within the jurisdiction in question; and/or

[68] Jack Goldsmith and Tim Wu, *Who Controls the Internet? Illusions of a Borderless World* (OUP 2008) 159 (emphasis added).

[69] Consider eg *Hugo Grotius on the Law of War and Peace*, ed Stephen C Neff (CUP 2012) 112.

- parties within the control of that government (eg residents or citizens) are not allowed to trade with the foreign party.

Parallels to such market destroying measures can be found in international relations. As seen, for example, in Ukraine, the world's response to the Russian invasion has so far been (mainly) in the form of sanctions we are used to in international relations. Thus, it is interesting to consider the role of sanctions in international relations. Sanctions, in that setting, are defined to mean:

> The use or threat of use of economic capacity by one international actor, or group of such actors against another international actor or group of actors, with the intention of (a) punishing the latter for its breach of a certain rule or (b) preventing it from infringing a rule which the party applying sanctions deems important.[70]

Or perhaps put even more potently, sanctions are 'a tool for coercing a target to change a course of action or forgo some future course of action by altering the cost-benefit calculus of a decision-maker to favor the preferred policy of the sender'.[71] Thus, sanctions under international relations are very similar in nature and aim to the market destroying measures I am discussing here, and in exploring such measures further we have a wealth of materials to draw from in the literature on sanctions. In fact, as noted by Brockman-Hawe: 'If there were a Top 10 List of topics most scrutinized by the international academic community, the issues of apposite objectives and justifications, not to mention the effectiveness, impact and suitable method of analysis of sanctions would probably turn up and occupy several spaces.'[72] In the light of options of the mentioned type of market destroying measures, the enforceability of jurisdictional claims—including those of a so-called 'extraterritorial' nature—may not be as limited as it may seem at a first glance. As hinted at earlier, we can usefully distinguish between the 'extraterritorial enforceability of the extraterritorial claim' as opposed to the 'domestic enforceability of the extraterritorial claim'. It is only the former that is suffering in efficacy, not the latter. This is a key point indeed, because it defeats the argument that extraterritorial claims lack means for enforcement—at least from a jurisprudential perspective, and possibly in practice too, we have here found some

[70] Sofie Heine-Ellison, 'The Impact and Effectiveness of Multilateral Economic Sanctions; A Comparative Study' (2001) 5 Intl J Hum Rights 81, 83.

[71] Jason C Nelson, 'The United Nations and the Employment of Sanctions as a Tool of International Statecraft: Social Power Theory as a Predictor of Threat Theory Utility' (2005) 29 Law & Psychol Rev 105, 111.

[72] Benjamin Brockman-Hawe, 'Using Internet "Borders" to Coerce or Punish: The DPRK as an Example of the Potential Utility of Internet Sanctions' (2007)25 BU Intl LJ 163, 182.

'protective footwear' that addresses and overcomes the Achilles' heel of extra-territoriality, albeit that the concept as such remains flawed.

6.5.1 Can Market Destroying Measures Delineate Market Sovereignty?

In this context, it is interesting to reconnect to the thinking of seventeenth-century legal scholars, exemplified by Hugo Grotius. Grotius stated:

> It seems clear, moreover, that sovereignty over a part of the sea is acquired in the same way as sovereignty elsewhere, that is, [. . .] through the instrumentality of persons and territory. It is gained through the instrumentality of persons if, for example, a fleet, which is an army afloat, is stationed at some point of the sea; by means of territory, in so far as those who sail over the part of the sea along the coast may be constrained from the land no less than if they should be upon the land itself.[73]

A similar reasoning can usefully be applied in relation to sovereignty in the context of the Internet. Instead of focusing on the location of persons, acts or physical things (as is traditionally done for jurisdictional purposes) we may perhaps turn our attention to marketplace control—to what we can call 'market sovereignty'. Perhaps we can argue that a state has market sovereignty and, therefore arguably justifiable jurisdiction, over Internet conduct where it can effectively exercise 'market destroying measures' over the market that the conduct relates to.[74] Thus, this doctrine of market sovereignty represents a move from jurisdiction based on perceived authority to jurisdiction based on actual control,[75] and importantly, emphasis is here placed on control of the relevant market rather than over the relevant conduct.[76]

[73] Consider eg *Hugo Grotius on the Law of War and Peace*, 112.

[74] The exercise of such measures must necessarily be carried out in a non-discriminatory manner and will doubtless give rise to interesting questions under eg the General Agreement on Trade in Services, and possibly under EU anti-discriminatory measures. However, those discussions, while interesting in the extreme, go beyond the scope here.

[75] Compare eg Nagan's discussion of how mechanisms such as 'long-arm' statutes represent a shift from 'control' to 'authority' (WP Nagan, 'Conflicts Theory in Conflict: A Systematic Appraisal of Traditional and Contemporary Theories' (1981–82) 3 J Intl & Comp L 343, 523).

[76] An example may be illustrative. A company in State A is selling goods to people in State B via the Internet. In such a scenario, State A may put an end to the company's activity by making it unlawful to sell such goods to State B. Leaving aside the obvious exception of situations where such a ban would violate some international agreement, this is uncontroversial. However, State B may also claim jurisdiction over the conduct by reference to its market sovereignty as it can potentially put an end to the company's activity by implementing market destroying measures.

Admittedly, allowing jurisdictional claims based on the doctrine of market sovereignty, put forward here, may give rise to unhelpfully wide jurisdictional claims. This could become problematic and may be counter-productive for the proposed aim of jurisdictional interoperability, not least in the light of claims that 'extraterritorial' law is inherently irreconcilable with democratic principles since such laws regulate foreigners who have not elected those people who made the law in question.[77] At the same time, where the violation of the law of a State merely results in such market destroying measures, a foreign defendant can comfortably carry on dealing with other markets should it not wish to comply with the laws of that State—importantly, the laws of one State do not then impose on foreign parties not interested in the marketplace to which that law applies, and if a party genuinely wishes to engage in the market in question, it can arguably not complain about having to comply with the same laws that apply to all other actors in that market.[78] Along a similar vein, it would seem peculiar for foreign States to complain about 'extraterritorial' claims of jurisdiction where such claims are only pursued in relation to market destroying measures; after all, the regulation in question is then strictly limited to parties engaging in the marketplace over which the lawmaker in question has sovereignty.

As to the claim of extraterritorial law's irreconcilability with democratic principles, a sober-minded consideration of the real state of things makes clear that such a claim cannot be maintained where the extraterritoriality is based on the doctrine of market sovereignty. After all, each party decides for itself which markets it wishes to act on, and the burden on, for example, a US company to abide by EU law when genuinely engaging in the EU market is not fundamentally different to the burden placed on an American tourist to abide by French law when visiting the Eiffel Tower.

Furthermore, at least in the online context, any party wishing to avoid the type of contact with, for example, the EU giving rise to a jurisdictional claim based on the doctrine of market sovereignty can do so by taking relatively

[77] For more on this (misguided) argument, see eg Austen Parrish, 'The Effects Test: Extraterritoriality's Fifth Business' (2008) 61(5) Vanderbilt LR 1456, 1483–89.

[78] It will clearly be necessary to define what is meant by a 'market'. However, guidance for such a venture could perhaps be gained from eg the field of antitrust law. One of the most important tasks in finding such a definition will be to avoid one that results in an overreach. For example, the definition of 'market' must avoid covering a situation where a person travelling to another country is seen to expand the market through her or his travels. In other words, the fact that a country can exercise market destroying measures by forbidding its nationals from entering into contract with certain businesses even while those nationals are overseas should not mean that the country in question has market sovereignty over the business.

simple technical steps to avoid such contact by 'dis-targeting' the EU.[79] Geo-location technologies have been available for quite some time now, and while the law has been slow to recognise their impact, businesses for example have been eager adopters (see Chapter 11).[80] Importantly, however, the European Commission has advanced a proposal for a Regulation[81] that would prevent the use of geo-location technologies in several types of situation. If adopted, this Regulation will have serious implications.

At any rate, this takes us to the question of recognition and enforcement by foreign courts. In most cases, the use of market destroying measures would make redundant the need for recognition and enforcement by foreign courts. However, where such a need exists, it will of course be for each state to decide whether or not to recognise and enforce a foreign judgment invoking market destroying measures on a party in that state. In many cases, doing so would perhaps not be objectionable as the effect is merely to prevent a particular party from entering a market overseas.[82] But in any case, we may have to abandon for now the idea that there should be sufficient harmony to allow overseas enforcement of all decisions in this field. Perhaps we simply are not there yet. Such surrender to reality should not be seen as a flaw in the international legal system. Instead it should be seen as a version, albeit a more primitive version, of the international system we should strive towards.

In the end, it is of course true that market sovereignty will often coincide with delineations founded on sovereignty based on persons or territories. However, conceptually the doctrine of market sovereignty is distinct—it has what could be viewed as a legal psycho-social focus to

[79] See further Dan Svantesson, 'Pammer and Hotel Alpenhof—ECJ Decision Creates Further Uncertainty about when E-businesses "Direct Activities" to a Consumer's State under the Brussels I Regulation' (June 2011) 27(3) Computer L & Sec Rev 298–304.

[80] See further Dan Svantesson, 'Time for the Law to Take Internet Geo-location Technologies Seriously' (December 2012) 8(3) J Priv Intl L 473–87; Dan Svantesson, 'Borders On, or Border Around—the Future of the Internet' (2006) 16(2) Albany LJ Sci & Tech 343–81; Dan Svantesson, 'Geo-Location Technologies and Other Means of Placing Borders on the "Borderless" Internet' (Fall 2004) XXIII(1) John Marshall J Compu & Info L 101–39; M Trimble, 'The Future of Cybertravel: Legal Implications of the Evasion of Geolocation' (2012) 22 Fordham Int Prop, Media & Ent LJ 567; KF King, 'Personal Jurisdiction, Internet Commerce, and Privacy: The Pervasive Legal Consequences of Modern Geolocation Technologies' (2011) 21 Albany LJ Sci & Tech 61, 66–67.

[81] Proposal for a Regulation of the European Parliament and of the Council on addressing geo-blocking and other forms of discrimination based on customers' nationality, place of residence or place of establishment within the internal market and amending Regulation (EC) No 2006/2004 and Directive 2009/22/EC, COM(2016) 289 final.

[82] One should, of course, acknowledge that a state has an interest in businesses established within the state benefiting from overseas markets.

spatial delineations—and has a much greater potential for future usefulness. Moreover, the doctrine of market sovereignty helps us move away from simple territoriality thinking.

It is also worth noting that the doctrine of market sovereignty highlights that the Internet does not force us to abandon the concept of sovereignty. This is significant as the obvious fundamental differences between various countries show that the world is far from ready to abandon sovereignty.

6.5.2 When is the Exercise of Market Sovereignty Appropriate?

If the arguments presented gain support, the question arises as to when a state should exercise market destroying measures based on market sovereignty. This question involves a difficult balancing of competing interests.

Where the law is faced with such matters of balancing, it often falls back on some form of proportionality assessment. Similarly, a proportionality assessment can usefully be applied to solve the question of when a state should exercise market destroying measures based on market sovereignty.

Here, a dispute involving Facebook in Brazil is illustrative. In the briefest terms possible, the case involved allegedly defamatory postings on Facebook, resulting in a Brazilian celebrity seeking the removal of those postings in Brazil. Facebook unsuccessfully fought the matter with the result that the Court ordered Facebook to remove the content within forty-eight hours. Failure to do so would result in Facebook being blocked in Brazil.

This case demonstrates that great care must be taken in identifying what values we are balancing in the proportionality assessment I advocate. Was the court order proportionate in the Brazilian case? The answer is 'no' if we are balancing the reputational harm against the societal harm of shutting down Facebook in Brazil. However, that comparison is misguided. We need to break down the situation into two parts. The first proportionality assessment should ask whether an order requesting the blocking of the defamatory content is proportionate to the reputational harm. I think most people would agree that it may be. In fact, this seems uncontroversial bearing in mind hundreds of years of legal tradition. The second, more complex, proportionality assessment relates to whether it is proportionate to block Facebook in Brazil if Facebook does not comply with the Court's order. What we are really comparing here is the societal harm of shutting down Facebook on the Brazilian market against the societal harm of concluding that society's need for Facebook is such that Facebook is *above* the law. Arguably, equal application of the law to all must triumph in any such assessment (however, as is discussed further in Chapter 12, a balanced solution must be found).

The remaining question then is whether the Court's threat could be put in other terms. What leverage does the Court have to make Facebook comply if it should not threaten to block Facebook? I argue that it could have started by the threat of other, less dramatic, market destroying measures.

In any case, the virtues of market destroying measures compared to traditional enforcement attempts could escape no one. Rather than interfering with the business operations worldwide in case of a dispute, market destroying measures only affect the offender's business on the market in question—it is thus a much more sophisticated and targeted approach. Where a globally active Internet intermediary finds compliance with a court order untenable, it will simply have to be prepared to abandon the market in question, but is free to pursue business elsewhere. Thus, an international agreement under which states undertake to only apply market destroying measures and not seek further enforcement would for example address the often excessive threat of arrests of key figures, such as CEOs, of offending globally active Internet intermediaries.

6.6 Concluding Remarks Regarding Functions of Jurisdictional Law

This chapter places the questions of Internet jurisdiction in a broader theoretical and, indeed philosophical, context. In addition, it provides distinctions important for how we understand the role of jurisdictional rules.

The tale of how King *Knut den store* (Canute the Great), ruler of England, Denmark, Norway, and parts of Sweden, commanded the waves to stop rolling is well known. And the outcome—the unsurprising fact that the waves disregarded his command—is equally well known. This has led to widespread ridicule of King Knut and the said event is frequently (mis)used as an example of foolish, arrogant, and futile attempts to prevent unstoppable forces.[83] Assuming the tale has any truth to it at all, the reality seems to be that King Knut did what he did in an attempt to demonstrate how limited his power was compared to that of the God people more commonly believed in at that time in Europe.[84]

[83] See eg John Perry Barlow, 'Jackboots on the Infobahn: Clipping the Wings of Freedom' (1993) 4 Wired <http://w2.eff.org/Misc/Publications/John_Perry_Barlow/HTML/infobahn_jackboots.html>, accessed 4 April 2017, bringing up 'the folly of King Canute' in the context of arguably futile attempts to stop the use of encryption.

[84] Kathryn Westcott, 'Is King Canute Misunderstood?' (BBC News, 26 May 2011) <http://www.bbc.co.uk/news/magazine-13524677>, accessed 4 April 2017. See also James Baldwin, 'King Canute on the Seashore' in William J Bennett (ed), *The Book of Virtues for Young People: A Treasury of Great Moral Stories* (Silver Burdett 1995).

It is no doubt tempting also to misapply the tale of King Knut to the context of jurisdictional claims over Internet activities—those who make such claims conscious of the unlikely enforceability (bark jurisdiction) may indeed be seen as brothers and sisters of King Knut in the misunderstood version of the tale. However, this chapter has sought to show that jurisdictional claims, even where their prospects of being enforced are dim, may have both jurisprudential legitimacy and practical utility.

The chapter also introduces and discusses the concept of 'market sovereignty' based on 'market destroying measures', noting how those concepts may be useful pieces in any attempt aimed at solving the Internet jurisdiction puzzle.

7

The Vagueness of the Law and the Importance of its Interpretation

While we often discuss the law as if it had a clear meaning, the reality is of course that most (if not all) application of the law involves the interpretation of the legal texts; that is, the meaning of the law can only be reached through the exercise of interpretation. This is particularly so in the Internet context,[1] and it is even more so if the theory of law I advanced in the previous chapter is accepted.

If we accept that law, as discussed in Chapter 6, is $n(Norm + {}_{Norm}(Context))$, then the law on a specific question x can be expressed as follows: $L_x = x(Norm + {}_{Norm}(Context))$ where L_x represents the law on that question, and $x(Norm + {}_{Norm}(Context))$ represents the legal norms and their context that are relevant for question x. Thus, due to the impossibility of ascertaining the full context, we can never conclusively ascertain what the law is in relation to a particular matter. To adopt Guastini's useful terminology, we are here dealing with the indeterminacy of any particular rule; that is, 'the law is indeterminate in the sense that it is not determinate what cases fall under the scope of each rule'.[2]

Put differently, the observation that we cannot ascertain the complete context of any particular legal norm leads to the necessary conclusion that we can ascertain the law neither theoretically nor practically. Since $L_x = x(Norm + {}_{Norm}(Context))$, and as ${}_{Norm}(Context)$ is unascertainable, the law L_x is never exact. It is not something that is, in the sense of, for example, $L_x = 4.13$. Rather, the law 'is' a more or less abstract range of options, such as $2 < L_x < 6$;[3] that is, the law on a particular matter is located within the borders set by the

[1] Michael N Schmitt and Liis Vihul, 'The Nature of International Law Cyber Norms' in Anna-Maria Osula and Henry Rõigas (eds), *International Cyber Norms: Legal, Policy & Industry Perspectives* (NATO CCD COE Publications 2016) 46.

[2] R Guastini, 'Rule-Scepticism Restated' in Leslie Green and Brian Leiter (eds), *Oxford Studies in Philosophy of Law*, vol 1 (OUP 2011) 144.

[3] The numbers here are merely used to represent the range of options. Thus, the numbers used obviously have no meaning or significance as such.

relevant legal norms and their respective contexts. In more detail, the fact that the range of options that make up the law is 'more or less abstract' could more accurately be expressed in the following: $A < L_x < B$ where $A \approx 2$ and $B \approx 6$. Thus, while I hereinafter refer to $2 < L_x < 6$, it must be remembered that the 2 and the 6 are approximate values that cannot be exactly ascertained.

Importantly, in the majority of cases, the relevant range of options is narrow, and only in so-called 'hard cases' will the adjudicator be presented with a real range of options. If it is conceded that the law can be expressed as, for example, $2 < L_x < 6$, it is necessary to explain what is found outside this range; what is found below 2 and above 6? Put simply, the answer is that outside that range we find mistakes and unreasonable interpretations of what the law is. Such mistakes and unreasonable interpretations will merely constitute oddities in the context of the legal norm in question.

To conclude this part of the argument, I suggest that the fact that the law is a more or less abstract range of options is both unsurprising and eminently useful for society. As argued by Burin in 1966: 'The modern state cannot do without administrative and judicial discretion. The complexity of present-day technology and of economic and social relationships, the struggles of unequal power groups, the rapidity and unpredictability of social change, these and other problems of society overtax the resources of legislative bodies and defy purely normative treatment by general law.'[4]

7.1 Why Interpretations Vary

As the law consists of a range of more or less abstract options, any application of the law involves subjective value judgements; by choosing how we approach ${}_{Norm}(Context)$, we may be able to reach a particular desired application of the law L_x. In other words, the reason one person concludes that L_x is L_1 while another person concludes that L_x is L_2 is found in the fact that they have differing perceptions of, and approaches to, ${}_{Norm}(Context)$.[5] In this way, the theoretical framework advanced in the previous section can be used to explain why different people, including judges, may end up applying one and the same norm in different manners.

In more detail, there are at least three likely causes why reasonable persons give different interpretations to the same legal norm. First, bearing in mind

[4] FS Burin, 'The Theory of the Rule of Law and the Structure of the Constitutional State' (1965–66) 15 Am U LR 313–40, 340.

[5] See also R Guastini (n 2) 148–49, referring to A Ross, *On Law and Justice* (Stevens & Sons 1958).

that it is not possible to know all aspects of the context, people work from an incomplete understanding and, thus, may have differing perceptions of what the context is as such. Second, even where they are in complete agreement as to the composition of the context, they may interpret the context in different ways. Finally, differing conclusions may be reached even where they agree on the composition of the context and interpret it in the same way because of different balancing and emphasising of the components of the context, and of course, different conclusions may also be reached due to different understandings of the facts of a particular case.

The likelihood of different people coming to different conclusions for any, or all, of these reasons increases with the legal and factual complexity of the case at hand. Thus, such differences of opinion are more likely in what is commonly termed 'hard cases' than in what is termed 'soft cases'. One important conclusion that can be drawn from the above is that subjectivity, in all its shapes, may influence how a person approaches a particular legal norm and its context. This conclusion applies to both legislators and courts and highlights a need for greater transparency in those legal processes that apply the law, and claim to announce what the law *is*. Thus, where the meaning of a legal norm L_x is determined to be L_1 rather than L_2, the interpretation of the relevant context of the legal norm ${}_{Norm}(Context)$, that caused the conclusion that $L_x = L_1$, must be made clear to the greatest extent possible. I have explored the need for transparency in some detail elsewhere.[6]

7.2 Principles to Govern Interpretation

What has been said in the previous section makes clear that judges interpreting and applying the law are making subjective choices amongst a more or less abstract range of options. In the light of this, one of the tools we can use in our work towards jurisdictional interoperability is found in defining the rules that govern the interpretation of the law. In other words, we need to re-examine our interpretation practice so that the interpretation process is guided by principles that lead us towards jurisdictional interoperability. I hasten to acknowledge that jurisdictional interoperability obviously is not the only goal of our legal system, and thus I am not suggesting that our interpretation practices ought to be guided *only* by the goal of jurisdictional

[6] Dan Svantesson, 'What is "Law", if "the Law" is Not Something That "Is"? A Modest Contribution to a Major Question' (September 2013) 26(3) Ratio Juris 456–85.

interoperability, but rather that the goal of jurisdictional interoperability must be *part of* our interpretation practices.

In my view, there are three main principles that ought to guide judges making subjective choices amongst a more or less abstract range of options: 'fairness' (defined here as the search for the most appropriate resolution of the dispute at hand), 'consequence focus' (defined here as the search for the option having the most favourable consequences for the future), and 'harmonisationalism' (defined here as the search for the option that is most clearly in line with international best practice). While different people will have different views as to what is 'fair', the need for a 'fair' resolution of legal disputes ought to be uncontroversial. Thus, apart from noting the fact that the judge's biases will determine what she regards as fair, I will not discuss this goal further here. Suffice it to note that this guiding principle fits well within the realist tradition: '[Realism] claims that judges should not act as if they are under a duty to apply the rules because they always have a choice to depart from them to a greater or lesser extent. Realists urge judges to use that freedom constructively to do justice rather than to apply inappropriate rules in a mechanical fashion.'[7] The second guiding principle I advance here—what I call 'consequence focus'—warrants a more detailed discussion. In my view, when they are making a subjective choice amongst a more or less abstract range of options, courts need to pay close attention to the likely future consequences of their choice. Such a focus is far too rare. However, examples can be found, not least in the Internet setting. In the *Lindqvist* case, the ECJ looked to the future consequences of the matter under consideration in noting how:

> If Article 25 of Directive 95/46 were interpreted to mean that there is 'transfer [of data] to a third country' every time that personal data are loaded onto an internet page, that transfer would necessarily be a transfer to all the third countries where there are the technical means needed to access the internet. The special regime provided for by Chapter IV of the directive would thus necessarily become a regime of general application, as regards operations on the internet. Thus, if the Commission found, pursuant to Article 25(4) of Directive 95/46, that even one third country did not ensure adequate protection, the Member States would be obliged to prevent any personal data being placed on the internet.[8]

As applied in the *Lindqvist* case, interpretation guided by a consequence focus serves primarily as what we, speaking bluntly, may refer to as a 'stupidity filter'. It tells us that, where the consequences of adopting a particular

[7] M Stokes, 'Formalism, Realism, and the Concept of Law' (1994) 13 L & Phil 115–59, 125.

[8] Case C-101/01, Criminal Proceedings against Bodil Lindvist, 6 November 2003, para 69.

interpretation would take us into the territory of absurdity, or at least would involve so far-reaching negative implications that they are clearly undesirable, the warning bells must ring. As noted by the Court in the *Google Warrant* case[9] discussed in Chapter 3: '[C]ourts should avoid an interpretation of a statute that produces odd or absurd results, or that is inconsistent with common sense.' Regrettably, there are several decisions on Internet jurisdiction that are more accurately described as 'uncommon nonsense' than as expressions of 'common sense'.

As soon as we go beyond this use of the consequence focus, we end up with a principle of interpretation that fits within the consequentialist thinking. And where that happens, we are exposed to vulnerabilities, for example as Kuner discusses in the context of more recent jurisprudence from the CJEU:

> The Court's current approach to defining the territorial scope of EU law on the Internet is largely based on the policy objectives that the law seeks to pursue. For example, the result of the Court's Google Spain judgment has been described as follows: 'the (territorial) scope of application of EU secondary law is determined by its policy objectives: a direct correlation can be established between the achievement of EU policies and the potential need to cover situations located in third states'. The greater the paucity of guidance by the legislator as to the territorial scope of law, the higher the risk that courts will be left to determine it based on their interpretation of the EU's policy objectives of the moment. Such a consequentialist approach risks sacrificing the coherence and consistency that are inherent in a legal (rather than a political) method of interpretation, and may result in the entanglement of legal values and political interests that has already been discussed.[10]

The consequence focus advocated here should not be confused with 'consequentialism'. Consequentialism—put simply, the end justifies the means—can be seen as the opposite of deontology which focuses on whether the conduct in question is 'right' or 'wrong'. What I am referring to in pointing to the need for consequence focus is unrelated to this debate and instead should be seen as the opposite of strict adherence to precedents.[11]

Thus, the consequence focus advocated here also fits well within the realist tradition and can be placed in contrast to a strict adherence to

[9] In re Search Warrants Nos 16-960-M-01 and 16-1061-M to Google, 26. In this context, the Court referred to: *Disabled in Action of PA v Southeastern PA Trans Auth*, 539 F3d 199, 210 (3d Cir 2008), *American Tobacco v Patterson* 456 US 63, 71 (1982), and *Register v PNC Financial Services Grp Inc* 477 F3d 56, 67 (3d Cir 2009).

[10] Christopher Kuner, The Internet and the Global Reach of EU Law (1 February 2017) Forthcoming in the Collected Courses of the Academy of European Law (OUP) 40 <https://ssrn.com/abstract=2890930>, accessed 2 April 2017 (footnote omitted).

[11] For an interesting discussion of consequentialism and deontology, see eg CS Man, 'Conflict in Canengus: The Battle of Consequentialism and Deontology' (2010) 7 J Juris 535–44, 535.

precedents, which represents a focus on history. There is no doubt that history is an important guide. However, the time has come to venture away from the steady decay caused by the current overreliance on past wisdom such as the 'original intention' and principles gained from outdated precedents (in both Common Law and Civil Law systems), and instead to look forward. In other words, the value of a decision taken today is better judged by the consequences it will have for the future than by how well it corresponds with what has been decided in the past—one cannot safely navigate forwards if one only looks in the rear-view mirror. Put bluntly, precedents ought to be used as a light to illuminate the correct path forward not as an anchor preventing the law from moving with the times.

The importance of harmonisationalism will only continue to increase with the escalating frequency and intensity of cross-border interactions. It will doubtless be a long process, but—as indicated in the discussion of legal interoperability—our aim must be to develop legal principles that are harmonised to an optimal degree. While important advancements towards this goal can be made through international agreements, such as treaties and model laws, national courts can also play an important role where they are guided by harmonisationalism.

To conclude this discussion, I acknowledge that the combination of fairness, consequence focus, and harmonisationalism arguably leads to less certainty than does a strict adherence to precedents. However, the law must always balance competing needs and, in my view, the benefits that can be gained from fairness, consequence focus, and harmonisationalism outweigh the negative impact of a lower degree of certainty.

7.3 A Common, and Particularly Unhelpful, Approach to Interpretation

Having outlined the principles that ought to guide our interpretation so as to cater for jurisdictional interoperability, it would not make much sense to devote space to discussing all interpretation practices that should be avoided. However, one such practice is both so common, and so unhelpful, that it deserves particular attention.

Perhaps the most significant cause contributing to making the system of international law dysfunctional in relation to matters of jurisdiction is found in the tendency of international law to be expressed and interpreted in overly absolute statements, better suited for consumption on the political arena than as expressions of legal norms. For example, in *Lotus*, Nyholm pointed to 'the principle of sovereignty and the territorial principle, accordance to which each nation has dominion over its territory and—on the other hand—has no authority to interfere in any way in matters taking place on the territories of other nations'.[12] A strict application of this absolutist statement is both non-sensical and totally unrealistic. After all, like the flap of a butterfly's wings in Brazil may set off a tornado in Texas, a product safety rule in Finland may cause a factory to be shut down in India, and seemingly a 'right to be forgotten' in the EU may determine Internet content availability worldwide. Or as put more elegantly by Michaels 'in a globalized economy, everything has an effect on everything'.[13]

An adherence to absolutist statements, such as Nyholm's, about the exclusiveness that flows from territorial sovereignty excludes the possibility of recognising, for example, objective territoriality, passive personality, or the effects doctrine. And yet we know that they are part of international law. On this point, Altamira, in his dissenting opinion in *Lotus*, made an important observation: 'It is not irrelevant to remark in passing that all or almost all principles of international law have the common characteristic of not being invariably exclusive in character.'[14]

7.4 Concluding Remarks Regarding Interpretation

The Finnish architect Eliel Saarinen instructed us always to 'design a thing by considering it in its next larger context—a chair in a room, a room in a house, a house in an environment, an environment in a city plan'.[15] This does not only apply to 'things'; it is also a useful instruction for lawmakers. Just about every decision made in law must be made with its broader context in mind. In defining—or in interpreting, as is discussed in this chapter—any legal term or concept, we must necessarily keep in mind that our decision affects a bigger picture. In the light of this, it is distinctly bad news when one sees a court suggest that it is engaging in 'a narrow question of statutory interpretation'.[16] There is simply no such thing as a *narrow*

[12] SS 'Lotus' (*France v Turkey*) (1927) PCIJ Series A, No 10, 59.

[13] R Michaels, 'Territorial Jurisdiction after Territoriality' in PJ Slot and M Bulterman (eds), *Globalisation and Jurisdiction* (Kluwer International 2004) 105, 123.

[14] SS 'Lotus' (*France v Turkey*) 102.

[15] <http://thinkexist.com/quotation/always_design_a_thing_by_considering_it_in_its/204270.html>, accessed 4 April 2017.

[16] A recent decision by the Federal Court of Australia is illustrative. In *Privacy Commissioner v Telstra Corporation Limited* [2017] FCAFC 4, the Court was essentially tasked with interpreting

question of statutory interpretation since that interpretation affects a bigger picture.[17]

Thus, as has been argued here, even where a legislator is providing definitions of terms used in legislation, and where a court is interpreting the meaning of a particular term, phrase or concept, the three interpretation principles discussed earlier must be adhered to.

Should it be the case that courts too often fail to appreciate the broader implications of their judgments in the Internet context, it may be time to abandon the long-lasting experiment of judicial ignorance and instead favour the development of specialised expert tribunals. This, however, is a huge topic of its own and need not detain us here.

the meaning of the words 'about an individual' in the context of a legal norm (National Privacy Principle 6.1) stating: 'If an organisation holds personal information *about an individual*, it must provide the individual with access to the information on request by the individual' (emphasis added). The Court was invited to consider the content of a detailed *amici curiae* brief <https://www.privacy.org.au/Papers/FCA-PCvT-160817.pdf>, accessed 4 April 2017, referring to a selection of relevant domestic and international materials. However, Kenny and Edelman JJ declined this opportunity as they saw themselves as merely tasked with a 'very narrow question of statutory construction' (*Privacy Commissioner v Telstra Corporation Limited* [2017] FCAFC 4 [73]).

[17] See also L Brilmayer, 'Liberalism, Community and State Borders' (1991) 41 Duke LJ 1, 1, noting: 'It should be of no surprise that even some of the humblest legal doctrines reflect much larger philosophical assumptions.'

8

The Impact of our Categorisation of Types of Jurisdiction

As already alluded to in the Introduction, convention would have us believe that jurisdictional issues arise in two distinct and neatly divided categories: private international law and public international law. As far as public international law goes, convention also suggests that jurisdiction may be equally neatly divided into three categories:[1] prescriptive (or legislative) jurisdiction,[2] adjudicative (or judicial) jurisdiction,[3] and finally enforcement jurisdiction.[4] And as we have learnt from the discussion of the 'Harvard Draft', prescriptive jurisdiction comes in various forms such as, for example, the territoriality principle, the nationality principle, and the universality principle.

The well-ordered picture painted so far is so common that the uninitiated may, indeed, be forgiven for thinking that it represents some form of truth—that it is the one and only 'correct' way to approach, structure, and understand jurisdiction. However, this is, of course, not so.

Let us start with the perceived distinction between, on the one hand, jurisdiction in private international law and, on the other, jurisdiction in public international law. We need not look far to find cracks in the uniformity of views. Just consider Ryngaert's observation that 'the law of jurisdiction

[1] See eg § 401 (a–c) *Restatement (Third) of US Foreign Relations Law*; Cedric Ryngaert, *Jurisdiction in International Law* (2nd edn, OUP 2015) 9–10; Michael N Schmitt (ed), *Tallinn Manual on the International Law Applicable to Cyber Warfare* (CUP 2013) 18.

[2] '[J]urisdiction to prescribe, i.e., to make its law applicable to the activities, relations, or status of persons, or the interests of persons in things, whether by legislation, by executive act or order, by administrative rule or regulation, or by determination of a court' (§401 (a) *Restatement (Third) of US Foreign Relations Law*). It may here be noted that this definition clearly supports the assertion that courts do make law.

[3] '[J]urisdiction to adjudicate, i.e., to subject persons or things to the process of its courts or administrative tribunals, whether in civil or in criminal proceedings, whether or not the state is a party to the proceedings' (ibid).

[4] '[J]urisdiction to enforce, i.e., to induce or compel compliance or punish noncompliance with its laws or regulations, whether through the courts or by use of executive, administrative, police, or other nonjudicial action' (ibid).

Solving the Internet Jurisdiction Puzzle. First Edition. Dan Jerker B. Svantesson.

bridges the private/public international law divide'.[5] Or why not consider Janis's 2016 textbook *International Law*. Chapter 10 is entitled 'International Conflicts of Laws'. Here, Janis explains that what he is describing is what Common Law states discuss as 'conflict of laws', and what civil law states discuss as 'private international law', and then proceeds to discuss the territoriality principle, the nationality principle, the effects principle, and other principles traditionally discussed as matters of public international law.[6]

The easy option would be to dismiss Janis's approach as a mistake. But even if we leave to one side Janis's extensive experience in the field and instead focus solely on the statement as such, would we really be justified in treating his approach as a mistake? Are we so sure about the 'correctness' of the conventional structure as to dismiss Janis's approach as misguided? I do not think so, and I think Janis goes some way towards explaining his unconventional approach by observing how:

> International law addresses not only the political and economic relations of nations, but also the interface between municipal legal systems. In civil law countries, this interface is studied under the rubric of *private international law* even though what is largely at issue are the international relations of courts, legislatures, and executives, surely a matter of public concern.[7]

At any rate, deciding whether Janis provides a plausible structure for how we approach jurisdiction is not necessary for our purposes here. Like others before me, I have for some time argued that there is an inevitable link between public and private international law.[8] Imagine, for example, that a Slovak court orders Google Inc to delist—with global effect—certain search results based on the Slovak defamation law. Imagine further that Google Inc complies with that order. What type of jurisdiction has the Slovak Republic then exercised? In answering this, some may be tempted to restrict themselves to pointing out that this is a matter of private international law and that we therefore do not need to try to fit the jurisdictional claim into the categories of legislative, adjudicative, and enforcement jurisdiction. But such an answer is lazy at best. After all, where a court makes a decision—whether in a

5 Ryngaert (n 1) 11; see also 16–22.

6 Mark Weston Janis, *International Law* (7th edn, Wolters Kluwer 2016) 333–41.

7 ibid 333.

8 Dan Svantesson, 'The Relation between Public International Law and Private International Law in the Internet Context' (2006) 9 Yearbook of New Zealand Jurisprudence Special Edition, 154–60. Consider also how certain fields, such as data privacy law, fall within both private and public international law (see further, Dan Svantesson, *Extraterritoriality in Data Privacy Law* (Ex Tuto 2013), and Ioannis Revolidis, 'Judicial Jurisdiction over Internet Privacy Violations and the GDPR: a Case of "Privacy Tourism"?' 11(1) Masaryk UJL & Tech (2017) 7.

civil, administrative, or criminal matter—it is the expression of the sovereign power of the court's state. And ultimately, the force of that decision is backed up by, indeed built upon, the enforcement powers of that state. As noted by Crawford: '[A]s civil jurisdiction is ultimately reinforced by criminal sanctions through contempt of court, there is in principle no great difference between the problems created by assertion of civil and criminal jurisdiction over aliens.'[9]

Turning to the categorisation within public international law, anyone bothering to examine the literature will soon find a serious lack of agreement, and it is clear beyond a reasonable doubt that not all commentators are praying at the same altar. Thus, we do well to acknowledge that the conventional structure is not the only plausible way in which we may approach matters of jurisdiction. Examples supporting this assertion abound.

Consider, for example, Hall's suggestion that '[t]here are *two* types of jurisdiction: prescriptive and enforcement [rather than three].'[10] In sharp contrast, we can note how Oster actually goes as far as to expressly extend the 'Harvard Draft' principles to adjudicative jurisdiction,[11] and how Putnam's interesting book *Courts Without Borders* is devoted to judicial jurisdiction[12]—precisely the type of jurisdiction Hall leaves out. Hall arguably also goes against convention in asserting that 'customary international law has developed rules and principles which define the limits of a state's *enforcement jurisdiction* by reference primarily to the state's territory, the nationality of the person affected by the exercise of jurisdiction, certain vital security interests of the state, and the defence of the international legal order as a whole'.[13] Further, in one passage, Crawford appears to equate the reach of enforcement jurisdiction with the reach of legislative jurisdiction: 'The present position is probably this: a state has enforcement jurisdiction abroad only to the extent necessary to enforce its legislative jurisdiction.'[14] Consider also how Putnam defines judicial jurisdiction to mean 'the practice of *domestic courts* unilaterally *applying domestic laws* to conduct and persons outside a state's borders',[15] while a more conventional definition would not necessarily focus on which country's laws are being applied in this context. The list could go on, but these selected examples ought to suffice to prove the point. To this we may add that, more

9 James Crawford, *Brownlie's Principles of Public International Law* (8th edn, OUP 2012) 471.

10 Stephen Hall, *Principles of International Law* (4th edn, LexisNexis 2014) 318 (emphasis added).

11 Jan Oster, *European and International Media Law* (CUP 2017) 115.

12 Tonya L Putnam, *Courts Without Borders: Law, Politics, and U.S. Extraterritoriality* (CUP 2016).

13 Hall (n 10) (emphasis added).

14 Crawford (n 9) 480.

15 Putnam (n 12) 1 (emphasis added).

than once, the complexity in distinguishing between the noted three types of jurisdiction has lead commentators into errors and lacking conceptual sharpness.[16] The resulting conflation of issues is harmful indeed.

At any rate, these 'idiosyncrasies' in how jurisdiction is classified under international law force us to concede that there is no uniform view on these matters. And in coming to accept this, we have made serious progress as we are then in a position to adopt a better structure without being accused of departing from *de lege lata*; all we need to do is express *de lege lata* in a better way.

Thus, as much as I would find it convenient to do so, I simply cannot subscribe to the widely held belief in the 'inviolate truth' of the structure described initially in this chapter. In addition to what has been noted already, several other severe obstacles prevent me from doing so. To start with a comparatively mild, but nevertheless arguably fatal, objection to the categorisation outlined, there is something distinctly unwholesome about a structure that gives a state a legitimate claim to have certain conduct governed by its criminal law but at the same time does not extend to that state the corresponding right to adjudicate a matter relating to that conduct, or indeed enforce that law against the offending party. Of course, proponents of the neat categorisation will hasten to point to the differing degrees of interference in other states' interest involved in the different types of jurisdictional claims. However, even where this is conceded, no one could point to this as a well-functioning structure. Furthermore, it does not always seem possible to fit disputes into the categories of legislative, adjudicative, and enforcement jurisdiction. A useful illustration is found in the discussed *Lotus* case. What type of jurisdiction—legislative, adjudicative, or enforcement—was that dispute about?

Amongst many others, Crawford, for example, approaches the *Lotus* case as a matter of prescriptive jurisdiction.[17] However, looking at the decision itself, things are not so clear as such commentators seem to suggest. Under the Special Agreement between the Government of the French Republic, and the Government of the Turkish Republic, the Court was to answer 'the question of *judicial jurisdiction* which has arisen between the two Governments'.[18] This is also reflected in the majority opinion observing that 'the proceedings relate exclusively to the question whether Turkey has or has not, according to

[16] See eg AK Woods, 'Against Data Exceptionalism' (2016) 68(4) Stan LR 729–90. For a detailed discussion of the problems with Woods's account of jurisdiction, see Dan Svantesson, 'Against "Against Data Exceptionalism"' (2016) 10(2) Masaryk UJL & Tech 200, 206–10.

[17] Crawford (n 9) 457.

[18] SS 'Lotus' (*France v Turkey*) (1927) PCIJ Series A, No 10, 21 (emphasis added).

the principles of international law, jurisdiction to prosecute in this case',[19] as well as in the dissenting opinions of Weiss.[20] At the same time, however, the picture is muddied somewhat by some of the dissenting opinions. Loder, for example, seems primarily focused on legislative jurisdiction: 'The fundamental consequence of their independence and sovereignty is that no municipal law, in the particular case under consideration no criminal law, can apply or have binding effect outside the national territory.'[21] Similarly, Nyholm focuses on legislative jurisdiction:

> [I]t cannot be maintained—as the judgment sets out—that, failing a positive restrictive rule, States leave other States free to edict their legislations as they think fit and to act accordingly, even when, in contravention of the principle of territoriality, they assume rights over foreign subjects for acts which the latter have committed abroad.[22]

Moore's dissenting opinion merges legislative and adjudicative jurisdiction into something governed by one principle:

> It is *an equally admitted principle* that, as municipal courts, the creatures of municipal law, derive their jurisdiction from that law, offences committed in the territorial jurisdiction of a nation may be tried and punished there according to the definitions and penalties of its municipal law, which, except so far as it may be shown to be contrary to international law, is accepted by international law as the law properly governing the case.[23]

Altamira also focuses on legislative jurisdiction combined with adjudicative jurisdiction.[24]

Furthermore, in discussing the fact that many states admit the principle of the exclusively and absolutely territorial character of criminal jurisdiction in their respective legislation, Weiss makes reference to how the French *Cour de cassation* declares in its judgment of 10 January 1873 (Dalloz, 1873, I 4I) that 'the right of inflicting punishment […] is derived from the right of sovereignty, which does not extend beyond the territorial limits'.[25] Surely this is a matter of enforcement jurisdiction. Thus, the judges in the *Lotus* case manage to approach the case as involving all three of the convention types of jurisdiction.

The inescapable conclusion is then that the Court in the *Lotus* case is inconsistent in its approach to what type of jurisdiction—legislative, adjudicative, and enforcement jurisdiction—to which the matter relates. Also, here, the easy option would be to dismiss this inconsistency as a mistake.

[19] ibid 13. See also ibid 12 and 15. [20] ibid 44.
[21] ibid 35. See also ibid 35–36 and 39. [22] ibid 60.
[23] ibid 68 (emphasis added). [24] ibid 102. [25] ibid 44.

However, as in the discussion at the beginning of the chapter relating to the unconventional approach to jurisdiction adopted by Janis, I suspect we would do ourselves a disservice if we merely dismissed this as a mistake. Rather, what we can learn from the inconsistency in the *Lotus* case is twofold; first, we ought to realise that it is not always possible, or indeed useful, to draw sharp lines between the different types of jurisdiction. And second, we are forced to accept that reasonable persons may view the proper categorisation differently. Both these conclusions undermine a strict adherence to the orthodox categorisation. In fact, having reached these conclusions, it would be illogical not to question our sleepwalking-like adherence to this categorisation of jurisdiction. After all, it is clearly both inadequate and prone to cause oversimplifications.

The first of these claims ought not need further support given the recent discussion. However, I acknowledge that I have not yet provided adequate support for the latter. I argue that we are at risk of relying on oversimplifications where we assume, for example, that claims of enforcement jurisdiction are always more intrusive—and therefore need to be strictly territorial—as compared to prescriptive jurisdiction which may be applied more broadly. Consider, for example, the situation where law enforcement in Portugal has a legitimate access to data stored, by the suspect, on a server in Singapore. Some would argue that in accessing that data, Portuguese law enforcement makes a claim of enforcement jurisdiction on Singaporean territory. But what, we may ask, is the degree of interference to which Singapore is subjected?[26] I would suggest that, in standard scenarios, this exercise of what traditionally would be seen as extraterritorial enforcement jurisdiction has very limited implications for Singapore, and indeed, no one in Singapore may even ever be aware that it was foreign law enforcement rather than the foreign account holder that accessed the data. In contrast, if Portugal were to prescribe that it is a crime for the people in Singapore to access, host, or share any information critical of the Portuguese national soccer team, this would amount to a significant interference with Singapore even without enforcement. Thus, in this comparison, the exercise of prescriptive jurisdiction causes a greater interference in a third country than does the exercise of what is typically seen as a form of enforcement jurisdiction. This is but one example, and the reader will have no difficulty thinking of others.

[26] For an insightful discussion of whether or not the state of the law enforcement agency (here Portugal) has a duty to notify the state in question where the data is held (here Singapore); see Anna-Maria Osula and Mark Zoetekouw, 'The Notification Requirement in Transborder Remote Search and Seizure: Domestic and International Law Perspectives' (2017) 11(1) Masaryk UJL & Tech 103.

In the light of the foregoing, I argue that the conventional focus on three (or two, as some suggest) distinct categories of jurisdiction—each associated with different rules for how broadly they may be applied—is both a source, and a sign, of confusion. And, thus, I am willing to go as far as to suggest that we may benefit from abandoning this categorisation. At the very least, we need to acknowledge that it does not represent the only way in which we may categorise jurisdiction.

Yet none the less, as its hold over the mind of many thinkers suggests, the categorisation of jurisdiction discussed here does contain, though in a blurred and misleading form, certain truths about certain important aspects of jurisdiction. When deconstructed, we see that what this categorisation is attempting to do is to bring attention to one single truth—the greater the potential a jurisdictional claim has of interfering with the sovereignty of another state, the greater the reason to contain, or limit, the exercise of jurisdiction. Or put differently, the degree to which a jurisdictional claim interferes with the sovereignty of another state must be part of the assessment of the reasonableness and legitimacy of that claim of jurisdiction. And this is a principle we may more easily apply in the absence of a misguided categorisation forcing us to pigeonhole jurisdictional claims as being either prescriptive jurisdiction, adjudicative jurisdiction, or enforcement jurisdiction. Importantly, this reasoning also lends support to the interest balancing that forms part of the jurisprudential framework for jurisdiction I have advanced earlier in this book.

8.1 Investigative Jurisdiction

If there is a value in continuing to categorise different types of jurisdiction, I propose that there is one category of jurisdiction that needs to be added: what I refer to as 'investigative jurisdiction'. Traditionally, investigative measures are—as already noted—viewed as an aspect of enforcement jurisdiction. Thus Crawford, for example, expresses a commonly held belief when he proclaims: 'Persons may not be arrested, a summons may not be served, police or tax investigations may not be mounted, orders for production of documents may not be executed, on the territory of another state, except under the terms of a treaty or other consent given.'[27] Conventional, yes, but how does he arrive at this insight? What sources back up the assumption that investigative measures must be treated as a part of enforcement jurisdiction?

Crawford points to three different authorities in the footnote he associates with the quoted text. The first is page 18 of the *Lotus* case. However,

[27] Crawford (n 9) 479.

neither the text on that page nor the *Lotus* case as a whole tells us anything about how we should categorise investigative measures. The second is a case summary of a decision from 1961 by the Austrian Supreme Court on the service of summons by post on a foreigner abroad.[28] And the third is a book—described in a book review published in the *American Journal of International Law*, as 'a rather interesting collection of cases and documents dealing with (a) treatment of aliens and (b) state territory, for the period 1914–1939'.[29]

While these authorities clearly hold some value, it would seem eccentric to suggest that they amount to a solid foundation upon which we may base the conclusion that investigative measures must necessarily be treated as a part of enforcement jurisdiction. In fact, doing so makes little sense—not least given that the online environment so clearly challenges the traditional link between location and control—and I would here like to try and dispute such an idea that I find quite unwholesome.

Imagine that State A sends agents into State B and kidnaps a citizen of State B so as to bring him before the courts in State A. Such matters fall within enforcement jurisdiction. Now imagine instead that law enforcement in State A has lawfully gained access to a laptop belonging to a citizen of State A. When they examine that laptop, they find that some relevant files are stored locally on the laptop while others are stored in 'the cloud' on servers in State B. Such a matter, although dramatically different to the first situation, would also—according to convention—fall within enforcement jurisdiction.

The problem is obvious: the policy considerations present in an assessment of jurisdiction in the latter scenario are fundamentally different from those involved in the former scenario. Or, to reconnect to what was advanced earlier, the degree to which a jurisdictional claim associated with kidnapping carried out on foreign soil interferes with the sovereignty of another state is much higher than the degree of interference caused by a jurisdictional claim associated with accessing files on a server located on foreign soil. Thus, the reasonableness and legitimacy of the claim of jurisdiction in the former scenario is much harder to establish than is the reasonableness and legitimacy of the claim of jurisdiction in the latter scenario. To then treat them as bound by the same jurisdictional rules is quite

[28] Service of Summons (1961) 38 ILR 133.

[29] Gerhard Von Glahn, 'Book Review: Répertoire suisse de droit international public: documentation concernant la pratique de la Confédération en matière de droit international public 1914–1939' (1978) 72 AJIL 197, 198.

simply indefensible. Surely we can all agree that, as a general rule, the threshold for jurisdiction must be higher in situations such as the kidnapping scenario compared to the threshold in the scenario involving law enforcement access to cross-border cloud data? The fact that international law is content to treat these situations as similar enough to be governed by the same rules is truly bizarre and reveals a troubling lack of refinement.

Support for the existence of a separate category of 'investigative jurisdiction' is increasing. For example, examining some recent cases decided by the CJEU, we can see hints at investigative jurisdiction being treated as a separate issue.[30] However, support can also be found outside Europe and in a case dating back to 2007. In *Lawson v Accusearch, Inc*, the Privacy Commissioner of Canada was forced to defend, in court, her decision to decline to investigate a complaint made by Lawson of the Canadian Internet Policy and Public Interest Clinic against a corporation based in the United States. Sean J Harrington of the Federal Court stated:

> I agree with her [the Privacy Commissioner of Canada] that PIPEDA [Personal Information Protection and Electronic Documents Act] gives no indication that Parliament intended to legislate extraterritorially. [. . .] [However, the] Commissioner does not lose her power to investigate because she can neither subpoena the organization nor enter its premises in Wyoming. [. . .] It would be most regrettable indeed if Parliament gave the Commissioner jurisdiction to investigate foreigners who have Canadian sources of information only if those organizations voluntarily name names. Furthermore, even if an order against a non-resident might be ineffective, the Commissioner could target the Canadian sources of information. I conclude as a matter of statutory interpretation that the Commissioner had jurisdiction to investigate, and that such an investigation was not contingent upon Parliament having legislated extraterritorially[.][31]

The instances where investigative jurisdiction plays a central role are numerous, for example, in the context of data privacy law and in areas such as consumer protection—areas where complaints are often pursued by bodies such as privacy commissioners/ombudsmen and consumer protection agencies. However, as is illustrated by the example just mentioned, its usefulness is rather more universal and clearly extends into, for example, the criminal law arena.

[30] See C-362/14 (*Schrems*) and C-230/14 (*Weltimmo*).

[31] *Lawson v Accusearch Inc dba Abika.com* [2007] 4 FCR 314 [28], [42], [43]. See also eg *Weltimmo s.r.o.v. Nemzeti Adatvédelmi és Információszabadság Hatóság* (Case C-230/14).

8.2 Jurisdiction over the Matter versus Jurisdiction over the Data

In connection with the discussion of 'investigative jurisdiction', another example of common conflation ought to be mentioned. Consider the following statement aimed at convincing readers that all is well in the application of existing jurisdictional principles to the context of cloud data:

> Jurisdiction is and likely always will be rooted in territoriality. States are the sovereigns of their territory and their citizens. Accordingly, they can regulate acts taking place on their soil as well as acts that affect their citizens, regardless of the location of those acts. This means that a state might legitimately assert its jurisdiction over a piece of data because that data or its controller is located in the state's territory, or simply because the data is needed for law enforcement there, regardless of where the data is stored or where the company is headquartered.[32]

This is rather a frightening claim as it would mean that any state could demand access to any piece of data it wanted, anywhere in the world, simply because the data is needed for law enforcement there—for example, in order to pursue human rights advocates, political dissidents, and others perceived as undesirables by that state.

However, that is not the aspect of the claim I wish to focus on here. The relevant feature of the quote is that it conflates two separate matters. Under orthodox thinking, States can indeed regulate acts taking place on their soil as well as acts that affect their citizens, regardless of the location of those acts. However, that does not tell us anything about jurisdiction over a piece of data that is stored outside the state by a company with no link to that state.

We must distinguish between, on the one hand, jurisdiction over the activity to be investigated and, on the other, jurisdiction over the evidence needed for the investigation. This distinction must always be kept in mind and is clearly articulated, for example, in the Budapest Convention.[33] Article 22, the provision that deals in general terms with the topic of jurisdiction, relates only to jurisdiction over the offences prescribed in the Cybercrime Convention (ie Arts 2 to 11) and does not govern jurisdiction over the evidence.

32 Woods (n 16) (footnote omitted).

33 Council of Europe Convention on Cybercrime (ETS No 185).

8.3 Concluding Remarks Regarding the Impact of our Categorisation of Types of Jurisdiction

This chapter has sought to draw attention to the fact that, while there is what may be described as a conventional view of how we ought to categorise jurisdictional claims, there is a considerable diversity of views with no real consensus discernible as to what the categories are, how they relate to each other, how they should be delineated, and what jurisdictional thresholds are associated with each category.

By pretending that the conventional view is the only correct view, we have been missing the opportunity properly to assess what lies at the core of the issue. I have here argued that the core principle to guide us may be expressed in the following:

> The greater the potential a jurisdictional claim has of interfering with the sovereignty of another state, the greater the reason to contain, or limit, the exercise of jurisdiction.

This ought to be quite uncontroversial. However, and this is important, I argue that this holds true across the categorisation forcing us to pigeonhole jurisdictional claims as being either prescriptive jurisdiction, adjudicative jurisdiction, or enforcement jurisdiction. Thus, the attempt to turn the assessment into a mechanical exercise of finding which category of jurisdiction we are dealing with is not only inelegant, but severely flawed.

In the end, it must be admitted that attempting to squeeze all types of jurisdictional claims into three categories is somewhat akin to placing all the animals of the world into three categories—the chosen level of granularity is inadequate. But if that is indeed how we continue to approach jurisdiction, I argue that what I here have described as 'investigative jurisdiction' is like the platypus: an oddity that does not fit well within any of the categories orthodoxy forces upon us.

Having said this, I acknowledge the truth in Niccolò Machiavelli's observation that '[t]here is nothing more difficult to take in hand, more perilous to conduct, or more uncertain in its success, than to take the lead in the introduction of a new order of things.'[34] Thus, I realise that change in the dominant assumptions, although flawed as they are, will require broad and sustained supported.

[34] <https://www.brainyquote.com/quotes/authors/n/niccolo_machiavelli.html>, accessed 8 April 2017.

9

Scope of (Remedial) Jurisdiction

When private international law lawyers speak of jurisdiction, focus is typically placed on so-called personal jurisdiction (jurisdiction *in personam*) and subject-matter jurisdiction. The former relates to the court's power to adjudicate matters directed against a particular party while the latter relates to, for example, the substantive areas of law in respect of which the courts may adjudicate matters. If we view personal jurisdiction and subject-matter jurisdiction as the first two dimensions of jurisdiction, there may be a third dimension—what we can call the 'scope of jurisdiction' or, perhaps even more precisely, 'scope of remedial jurisdiction'.[1] Scope of jurisdiction relates to the appropriate geographical scope of orders rendered by a court that has personal jurisdiction and subject-matter jurisdiction. This question has gained surprisingly little attention to date. However, it is a question that is increasingly important—not least given the current hyper-regulation discussed in Chapter 4—and therefore deserving of detailed attention.

At a first glance, it is possible to form the view that scope of (remedial) jurisdiction is in fact not a separate matter at all. However, a finding that a court is entitled to claim *in personam* jurisdiction over a party and subject-matter jurisdiction over the type of dispute, and that the dispute falls within the relevant court's territorial competence and territorial jurisdiction does not tell us anything about the appropriate geographical scope of any orders such a court may render. Clearly then, scope of (remedial) jurisdiction is a separate matter.[2] Thus, while this third dimension is often overlooked, there can be no sensible dispute as to its existence as a separate issue, and clear examples can be found of its articulation. One such example is found in the following

[1] The latter term—'scope of remedial jurisdiction'—was used by Justice Groberman in a recent decision of the Court of Appeal for British Columbia (*Equustek Solutions Inc v Google Inc* [2015] BCCA 265 [69] <http://canlii.ca/t/gjgwv>, accessed 4 April 2017). However, I will predominantly use the shorter term, 'scope of jurisdiction'.

[2] See further Dan Svantesson, 'Jurisdiction in 3D—"Scope of (Remedial) Jurisdiction" as a Third Dimension of Jurisdiction' (2016) 12(1) J Priv Intl L 60–76, 62–63, 67–71.

Solving the Internet Jurisdiction Puzzle. First Edition. Dan Jerker B. Svantesson.

quote from the *eDate* case, commenting on the EU's Brussels Regulation's approach to jurisdiction in torts cases:

> Article 5(3) of the Regulation must be interpreted as meaning that, in the event of an alleged infringement of personality rights by means of content placed online on an internet website, the person who considers that his rights have been infringed has the option of bringing an action for liability, in respect of all the damage caused, either before the courts of the Member State in which the publisher of that content is established or before the courts of the Member State in which the centre of his interests is based. That person may also, instead of an action for liability in respect of all the damage caused, bring his action before the courts of each Member State in the territory of which content placed online is or has been accessible. Those courts have jurisdiction only in respect of the damage caused in the territory of the Member State of the court seised.[3]

This statement outlines rules of personal jurisdiction in stating, for example, that an action can be taken *before the courts of the Member State in which the publisher of that content is established* or *before the courts of the Member State in which the centre of his interests is based*. But it clearly and indisputably also delineates the scope of jurisdiction. For example, it makes clear that, if you sue at the defendant's place of establishment or where you have your centre of interests, the relevant scope of jurisdiction is global, while if you sue in a different country, the relevant scope of jurisdiction is just that Member State.

Admittedly, such clear articulations of scope of jurisdiction are the exception rather than the rule. That does not, however, amount to a reason to doubt that scope of jurisdiction issues exist and are important. Sometimes scope of jurisdiction issues are dealt with in a less direct, but perhaps equally effective, manner. Having mentioned the EU approach, we can, for example, make an interesting comparison with the practical effect the (Common Law) doctrine of *forum non conveniens* has on scope of jurisdiction.

Litigants who pursue global remedies, for example in a defamation matter, risk the court concluding that the matter is more appropriately dealt with in another court since such actions inevitably have links to more than one state. Thus, the net effect of the doctrine of *forum non conveniens* has often been to encourage the litigants to limit their claims to damages occurring in the state where the litigation takes place—the litigants voluntarily limit the scope of jurisdiction by suing only in relation to local damages and undertaking not to sue elsewhere so as to avoid the doctrine of *forum non conveniens*. The well-known *Gutnick* case[4] from Australia is a classic

[3] Cases C-509/09 *eDate Advertising GmbH v X* and C-161/10 *Olivier Martinez and Robert Martinez v MGN Limited*, para 52.

[4] *Dow Jones & Company Inc v Gutnick* [2002] HCA 56.

example of this, and the 2012 Canadian case of *Breeden v Black*[5] is a more recent example. Thus, in a sense, the doctrine of *forum non conveniens* fulfils the same function, in relation to scope of jurisdiction, as does the EU law framework alluded to already.

9.1 An Important Battleground

The issue of scope of jurisdiction must always have mattered. Indeed, the concept itself, and some sources of limitations to be placed on scope of jurisdiction, can perhaps be said to be hinted at in the following passage by Altamira in the *Lotus* case:

> Of course, every sovereign State may by virtue of its sovereignty legislate as it wishes within the limits of its own territory; but it cannot, according to sound principles of law, in so doing impose its laws upon foreigners in every case and without making any distinction between the various possible circumstances as regards the place where the offence has been committed, the nature and seriousness of the offence, the special conditions under which a foreigner may happen at a given moment to be within reach of the authorities of a foreign country on the territory of which the offence of which he has been accused was not committed, and other conditions besides.[6]

Furthermore, concern about the excessively broad scope of remedial jurisdiction has influenced the outcome in at least one early Internet jurisdiction case. In *Macquarie Bank Limited & Anor v Berg*, the Supreme Court of New South Wales expressed the view that:

> [a]n injunction to restrain defamation in NSW [New South Wales] is designed to ensure compliance with the laws of NSW, and to protect the rights of plaintiffs, as those rights are defined by the law of NSW. Such an injunction is not designed to superimpose the law of NSW relating to defamation on every other state, territory and country of the world. Yet that would be the effect of an order restraining publication on the Internet.[7]

In this we can see that, just like courts can extend their scope of jurisdiction too far, a fear of the scope of jurisdiction being too broad may also be harmful in that courts may decline to exercise personal jurisdiction due to the scope of jurisdiction perceived to follow, as a necessity, from such a claim.

5 [2012] SCC 19. See in particular [33].
6 SS 'Lotus' (*France v Turkey*) (1927) PCIJ Series A, No 10, 101–02.
7 *Macquarie Bank Limited & Anor v Berg* [1999] NSWSC 526 [14].

Be that as it may, there can be no doubt that the question is becoming increasingly important; indeed, it is now a centrally important battleground in the Internet jurisdiction context. The reason is obvious. Issues of scope of jurisdiction are particularly controversial in the context of removal, blocking, or delisting of Internet content since in that context there are substantial cross-border implications—a court's order in State A that content be removed from the Internet directly affects what content people in other states can access online.

This is so, not least as we see more and more litigation aimed at Internet intermediaries such as search engines, social media operators, and video hosting platforms. Such intermediaries have the ability to achieve global removal, blocking, or delisting of content based on that content being contrary to the law of one single state. Put dramatically, in directing lawsuits at such middlemen, litigants can 'kill a mosquito with a nuclear bomb'—local removal is ensured by global removal; and in this context, as also is the case more broadly with jurisdictional claims, states are showing preciously little 'jurisdictional self-restraint'.[8]

There are an increasing number of cases in which Internet intermediaries have been ordered to remove, block, or delist content based on the fact that the content in question violates the law of one specific state. In some such cases, the courts order global blocking without much discussion of the issue.[9] In other cases, the scope of jurisdiction matter is a central component in the dispute. Here, I will provide some details of the two most significant legal disputes to date that deal directly with the issue of scope of (remedial) jurisdiction.

9.1.1 The Aftermath of *Google Spain* (*González*)

Few Internet-related cases can compete in fame (or notoriety) with the CJEU's decision in the *Google Spain*—or 'right to be forgotten'—case.[10] Essentially what happened there was that, based on a complaint made by a Spanish man about the search results that were displayed if anyone searched his name on the Google search engine, the CJEU upheld a right for people to have certain search results delisted from search engines. However, the Court

[8] For an interesting discussion of 'jurisdictional self-restraint' see Graham Smith, 'Cyberborders and the Right to Travel in Cyberspace' in Uta Kohl (ed), *The Net and the Nation State—Multidisciplinary Perspectives on Internet Governance* (CUP 2017) 125–44.

[9] *Garcia v Google Inc* No 12-57302 (9th Cir 2014) <http://cdn.ca9.uscourts.gov/datastore/general/2014/02/28/12-57302_opinion.pdf>, accessed 4 April 2017. An appeal overturned the decision on the basis of copyright law.

[10] C-131/12.

was never asked to deal with the scope of jurisdiction question. As a direct, and unfortunate, consequence of this, there is considerable controversy about how widely—geographically speaking—search engines need to delist search results based on the so-called 'right to be forgotten'.

The case itself is illustrative of some of the available options. Should the delisting of the search results that appear upon a search of that man's name be applied only to Google's Spanish website 'google.es'? Or should perhaps the search results also be delisted on all other European domains like 'google.de', 'google.fr', and 'google.se'? Or should in fact the delisting apply also to the international site 'google.com'? Or even more broadly, to distinctly non-European domains like 'google.ca' for Canada, or 'google.com.au' for Australia? This matter is clearly a question of scope of jurisdiction, and it is no doubt the most hotly contested scope of jurisdiction question of today.

Google's initial implementation was described by its Global Privacy Counsel, Peter Fleischer:

> We do not read the decision by the Court of Justice of the European Union (CJEU) in the case C-131/12 (the 'Decision') as global in reach—it was an application of European law that applies to services offered to Europeans.
>
> [...] In its decision, the CJEU presented a legal interpretation affecting multiple countries simultaneously. We heard some DPAs [Data Protection Authorities] and others call for consistency across states in implementing it, and we have therefore decided to respect that effort by extending each removal to all EU/EFTA ccTLDs.[11]

In sharp contrast, the Article 29 Working Party's[12] 'Guidelines'[13] regarding the *Google Spain* decision emphasises that:

[11] Peter Fleischer, 'Response to the Questionnaire addressed to Search Engines by the Article 29 Working Party regarding the implementation of the CJEU judgment on the "right to be forgotten"' (31 July 2014) <https://docs.google.com/file/d/0B8syaai6SSfiT0EwRUFyOENqR3M/view?pli=1&sle=true>, accessed 4 April 2017.

[12] 'The Article 29 Data Protection Working Party is composed of: a representative of the supervisory authority (ies) designated by each EU country; a representative of the authority (ies) established for the EU institutions and bodies; a representative of the European Commission.' It was 'set up under the Directive 95/46/EC of the European Parliament and of the Council of 24 October 1995 on the protection of individuals with regard to the processing of personal data and on the free movement of such data. It has advisory status and acts independently.' European Commission, 'Article 29 Working Party' (11 September 2015) <http://ec.europa.eu/justice/data-protection/article-29/index_en.htm>, accessed 4 April 2017.

[13] Those Guidelines were 'designed to provide information as to how the DPAs assembled in the Working Party intend to implement the judgment of the CJEU in the case of "Google Spain SL and Google Inc v Agencia Española de Protección de Datos (AEPD) and Mario Costeja González" (C-131/12)': Art 29 Data Protection Working Party, 'Guidelines on the implementation of the Court of Justice of the European Union judgment on "Google Spain and Inc v Agencia Española de Protección de Datos (AEPD) and Mario Costeja González" C-131/12' (2014) WP225, 5.

de-listing decisions must be implemented in a way that guarantees the effective and complete protection of these rights and that EU law cannot be easily circumvented. In that sense, limiting de-listing to EU domains on the grounds that users tend to access search engines via their national domains cannot be considered a sufficient means to satisfactorily guarantee the rights of data subjects according to the judgment. In practice, this means that in any case de-listing should also be effective on all relevant domains, including .com.[14]

Thus, at least one message is clear: the Article 29 Working Party wants global blocking so as to ensure that EU law is not 'circumvented'. The question is, of course, whether blocking on the .com domain will be enough to achieve this. Or does the Article 29 Working Party actually intend search engines to impose delisting also for non-European country domains? Perhaps the Article 29 Working Party's wording gives us some hints. It emphasises that '[i]n practice, this means that in any case de-listing should also be effective on *all relevant domains*, including .com' (emphasis added). If the intention was that delisting should apply to *all* domains, there would be no need for the word 'relevant'.[15] This suggests that what the Article 29 Working Party has in mind is delisting on EU domains and on the .com domain.

On the other hand, the very reasoning behind delisting on the international .com domain is that it is currently easy for people to use google.com to access content delisted on a country-specific search such as google.es—the Spanish domain. But, if content is delisted also on google.com, will not people who are sufficiently motivated to search for the content simply use another non-EU country-specific search such as google.com.au? After all, doing so requires little extra effort.

Will this reasoning then mean that, to comply with EU law, search engines need to delist search results all over the world, including on distinctly non-EU domains such as Australia's .com.au, Japan's .jp and Colombia's .co? Given the Article 29 Group's decision to issue 'Guidelines', it would have been useful if it would have engaged with this topic of fundamental importance more thoroughly and more thoughtfully.

In any case, in a media release of 12 June 2015, the French data protection authority—the Commission Nationale de l'Informatique et des Libertés (CNIL)—stated, amongst other things, that:

CNIL considers that in order to be effective, delisting must be carried out on all extensions of the search engine and that the service provided by Google search

[14] ibid 9.

[15] While far-fetched, it is of course possible that the word 'relevant' here is aimed to clarify that it is the domains belonging to a specific search engine that are referred to. However, that would be an unnecessarily cryptic way to express such delineation.

constitutes a single processing. In this context, the President of the CNIL has put Google on notice to proceed, within a period of fifteen (15) days, to the requested delisting on the whole data processing and thus on all extensions of the search engine.[16]

This gave a clear indication that the dispute would carry on long after the CJEU's original decision, and this saga is ongoing at the time of writing.[17] That, however, has not prevented courts in other parts of the world being quick to embrace the CJEU's findings in the *Google Spain* case. For example, the lower court in the *Google Canada*[18] case discussed in the following section, relied expressly on the CJEU's decision as an authority for linking Google Inc to the activities of Google's country-specific branches. However, in that context, it should be noted that, in a recent Opinion, Advocate General Saugmandsgaard Øe has emphasised the special nature of how the EU Data Protection Directive applies to that case in noting:

> [T]hat case [*Google Spain*] differs from the present case in that it was a matter of assessing, in that case, whether the processing of data concerned was covered by the framework for protection established by Directive 95/46 (through the Spanish law transposing it). It was, in my view, from that perspective that the Court interpreted broadly the second condition laid down in Article 4(1)(a) of that directive in order to prevent such processing from escaping the obligations and guarantees provided for in the directive.[19]

Advocate General Saugmandsgaard Øe's remarks in this respect were not discussed by the CJEU in the matter the Advocate General was commenting on, and nothing was said that contradicts the accuracy of his logical claim.[20] Thus, care must be exercised in what is read into the *Google Spain* case, and

[16] 'CNIL orders Google to apply delisting on all domain names of the search engine' (CNIL, 12 June 2015) <http://www.cnil.fr/english/news-and-events/news/article/cnil-orders-google-to-apply-delisting-on-all-domain-names-of-the-search-engine/>, accessed 2 April 2017.

[17] See eg Christopher Kuner, The Internet and the Global Reach of EU Law (1 February 2017) forthcoming in the Collected Courses of the Academy of European Law (OUP) 25 <https://ssrn.com/abstract=2890930>, accessed 2 April 2017, and Daphne Keller, 'Global Right to Be Forgotten Delisting: Why CNIL is Wrong' (November 18, 2016) <https://cyberlaw.stanford.edu/blog/2016/11/global-right-be-forgotten-delisting-why-cnil-wrong>, accessed 11 April 2017. Indeed, it is now clear that the CJEU will be asked to clarify several aspects of the 'right to be forgotten': 'Right to be delisted' (Conseil d'État, 24 February 2017) <http://english.conseil-etat.fr/Activities/Press-releases/Right-to-be-delisted>, accessed 2 April 2017.

[18] *Equustek Solutions Inc v Jack* (2014) BCSC 1063 <http://www.courts.gov.bc.ca/jdb-txt/SC/14/10/2014BCSC1063.htm>, accessed 2 April 2017.

[19] Opinion by Advocate General Henrik Saugmandsgaard Øe in *Verein für Konsumenteninformation*, Court of Justice of the European Union, C-191/15, 2 June 2016, para 124.

[20] Case C-191/15.

it must be seen to be somewhat careless for the Supreme Court of Canada to embrace the lower courts' reference to the *Google Spain* case without paying attention to the more recent devlopments I alluded to here.[21] However, one thing is undeniable—for better or worse, the *Google Spain* case has done much to put scope of jurisdiction on the agenda.

9.1.2 *Google Canada* (*Equustek*)

The second case illustration I want to point to in the context of scope of jurisdiction is the long-running dispute between Google Inc and Equustek Solutions Inc. The case has gained a considerable degree of international attention and the background to the dispute is generally well known. The issue in the appeal was whether Google could be ordered (via an interlocutory injunction) to de-index (with global effect) the websites of a company (Datalink Technology Gateways Inc, and Datalink Technologies Gateways LLC) which, in breach of several court orders, sell the intellectual property of another company (Equustek Solutions Inc) via those websites.

At first instance,[22] Justice Fenlon of the Supreme Court of British Columbia granted an injunction requiring Google to block certain websites worldwide. Google argued that if the injunction was granted it should be limited to 'google.ca', the website designated for Canada, for the reason that no court should make an order that has a reach that extends around the world; after all, what is illegal in State A may be perfectly legal, or even demanded by the law, in State B. However, the Court stated:

> [A]lthough Google has a website for each country to which searches made within that country default, users can override that default and access other country's Google websites. For example, even if the defendants' websites were blocked from searches conducted through www.google.ca, Canadian users can go to www.google.co.uk or www.google.fr and obtain results including the defendants' websites. On the record before me *it appears that to be effective, even within Canada, Google must block search results on all of its websites. Furthermore, the defendants' sales originate primarily in other countries, so the Court's process cannot be protected unless the injunction ensures that searchers from any jurisdiction do not find the defendants' websites.*[23]

Given the availability of alternative search engines, this argument is difficult to maintain. The fact that one particular search engine blocks certain search results worldwide does not affect the availability of the content in

[21] *Google Inc v Equustek Solutions Inc* 2017 SCC 34, para 39.

[22] *Equustek Solutions Inc v Jack* (2014) BCSC 1063 <http://www.courts.gov.bc.ca/jdb-txt/SC/14/10/2014BCSC1063.htm>, accessed 2 April 2017.

[23] ibid [148] (emphasis added).

question—the content is still there, it just cannot be found using that particular search engine. Consequently, the content can still be accessed through the use of an alternative search engine. But the Court, to its credit, had thought of this as well:

> While there are other search engines, Google does not contest the plaintiffs' assertion that Google's position as the search engine used for 70–75% of internet searches means the defendants will not be commercially successful if they cannot be found through Google's search services.[24]

With this in mind, the Court concluded that: 'Google is an innocent bystander but it is unwittingly facilitating the defendants' ongoing breaches of this Court's orders. There is no other practical way for the defendants' website sales to be stopped.'[25] The last sentence in this quote is particularly telling and, of course, incorrect. For example, sales require some form of payment method, so one other practical way for the defendants' website sales to be stopped would be to seek to cut the payment mechanism.

In any case, the court order was appealed and leave was granted.[26] However, Justice Groberman, writing the judgment for the Court of Appeal for British Columbia, came to much the same conclusions as did Justice Fenlon, and the Court of Appeal dismissed the appeal.[27] The matter was then appealed to the Supreme Court of Canada. That Court heard the matter on 6 December 2016, and on 28 June 2017 the Supreme Court of Canada handed down its judgment.[28]

The majority of the Court (McLachlin CJ and Abella, Moldaver, Karakatsanis, Wagner, Gascon, and Brown JJ) ruled in favour of Equustek concluding that:

> [S]ince the interlocutory injunction is the only effective way to mitigate the harm to Equustek pending the resolution of the underlying litigation, the only way, in fact, to preserve Equustek itself pending the resolution of the underlying litigation, and since any countervailing harm to Google is minimal to non-existent, the interlocutory injunction should be upheld.[29]

This conclusion was reached via a belaboured journey through the quagmire of both legal and technical misunderstandings and half-truths. Most of those misunderstandings and half-truths are highlighted with

[24] ibid [152]. [25] ibid [156].

[26] *Equustek Solutions Inc v Google Inc* (2014) BCCA 295 <http://www.courts.gov.bc.ca/jdb-txt/CA/14/02/2014BCCA0295.htm>, accessed 2 April 2017.

[27] *Equustek Solutions Inc v Google Inc* [2015] BCCA 265, [69] <http://canlii.ca/t/gjgwv>, accessed 2 April 2017.

[28] *Google Inc v Equustek Solutions Inc* 2017 SCC 34. [29] ibid para 53.

commendable clarity in the dissenting judgment by Côté and Rowe JJ who also stressed that 'In our view, granting the Google Order requires changes to settled practice that are not warranted in this case: neither the test for an interlocutory nor a permanent injunction has been met; court supervision is required; the order has not been shown to be effective; and alternative remedies are available.'[30]

Fenlon J,[31] the Court of Appeal,[32] and the Supreme Court of Canada discussed the issue of scope of (remedial) jurisdiction at various places, and pointed to versions of the standard test for granting interlocutory injunctions,[33] with the majority judgment of the Supreme Court of Canada concluding: "Ultimately, the question is whether granting the injunction would be just and equitable in all the circumstances of the case.'[34]

Regrettably, Fenlon J, the Court of Appeal, and the Supreme Court of Canada all failed to approach the issue of scope of (remedial) jurisdiction in a coherent manner recognising it as a distinct issue. Furthermore, they failed to recognise all the factors that ought to influence a court's decision as to the appropriate scope of (remedial) jurisdiction in the matter before it. It is particularly troubling that they failed to appreciate the extent to which the strength of the connection to the forum affects the appropriate scope of (remedial) jurisdiction.

9.2 A Coherent Framework for Scope of (Remedial) Jurisdiction

The decisions in the mentioned *Google Canada* case provide an excellent starting point for a discussion of the delineation of scope of (remedial) jurisdiction. So what needs to be considered if we are to devise a coherent framework for scope of (remedial) jurisdiction?

Given that a broad scope of (remedial) jurisdiction may significantly impact foreign countries and persons in foreign countries, a court determining the scope of (remedial) jurisdiction in a particular matter is bound by

[30] ibid para 60.

[31] *Equustek Solutions Inc v Jack* (2014) BCSC 1063 <http://www.courts.gov.bc.ca/jdb-txt/SC/14/10/2014BCSC1063.htm>, accessed 2 April 2017, in particular at paras 104, 108, 146, 150, and 159.

[32] Decision of the Court of Appeal of British Columbia dated 11 June 2015, paras 25, 60, 69, 85, 88, and 106.

[33] Reasons for Judgment of the Honourable Justice Fenlon dated 13 June 2014 [150], and decision of the Court of Appeal of British Columbia dated 11 June 2015 [100], respectively.

[34] *Google Inc v Equustek Solutions Inc* 2017 SCC 34, para 1.

both domestic (conflict of laws rules, its law of remedies, and constitutional law) and international law. This is uncontroversial and indeed is reflected in the judgments by the courts in the *Google Canada* matter.[35] In the light of the fact that the scope of (remedial) jurisdiction analysis must depend on both domestic and international law, I argue that at least five factors are relevant. I discuss them here expressed as a proposed comprehensive framework for dealing with scope of (remedial) jurisdiction.

The framework, and its five factors that must be considered by a court seeking to determine the legitimate and appropriate scope of (remedial) jurisdiction in any given matter, was first presented as part of my failed application for leave to intervene as *amicus curiae* when the matter came before the Supreme Court of Canada.[36] The five factors are:

1. the strength of the connection to the forum;
2. the connection between the party and the dispute;
3. the impact on other countries and persons in other countries;
4. the type and effectiveness of the order compared to alternatives; and
5. the level of fault of the party.

While the appropriate scope of jurisdiction is highly context-dependent, I argue that where account is taken of all these factors we have a sufficiently flexible framework to approach scope of (remedial) jurisdiction as one issue rather than approaching it in a sectoral sense.

9.2.1 The Strength of the Connection to the Forum

The legitimacy of a broad scope of (remedial) jurisdiction (such as the worldwide order sought in the *Google Canada* case) increases with the strength of the connection between the forum and the dispute and the parties as assessed for example in the analysis of *in personam* jurisdiction, subject-matter jurisdiction, and territorial competence. In general terms, a court has greater legitimacy in granting a worldwide injunction in a domestic dispute between two domestic parties than it has in making an order against a foreign party in an international dispute.

The lower courts—in the *Google Canada* case—failed to appreciate this connection between, on the one hand, *in personam* jurisdiction, subject-matter jurisdiction, and territorial competence, and on the other scope of

[35] See eg Decision of the Court of Appeal of British Columbia dated 11 June 2015 [88].

[36] <http://www.scc-csc.ca/case-dossier/info/dock-regi-eng.aspx?cas=36602>, accessed June 2017.

(remedial) jurisdiction. For example, Fenlon J asserts: 'If the order involves worldwide relief, a California court will be no more appropriate a forum than British Columbia to make such an order.'[37] But clearly, a California court has a stronger claim of personal jurisdiction over Google Inc than does British Columbia.

More broadly, in the light of how the strength of the connection between the forum and the dispute and the parties impact the legitimacy of the scope of (remedial) jurisdiction, it is not appropriate to 'compartmentalise' the reasoning in the manner done by the courts. A finding that the grounds for personal jurisdiction are only barely met ought to cause a court to hesitate to claim a broad scope of jurisdiction. For example, findings such as that the jurisdictional ground provided for under s 10(i) of the *Court Jurisdiction and Proceedings Transfer Act* (*CJPTA)* 'is not applicable',[38] that s 10(a) of the *CJPTA* 'may be a weak connecting factor'[39] and that the jurisdictional ground provided for under s 10(h) of the *CJPTA* is only described as 'stronger'[40] than the weak s 10(a) must necessarily carry over and also colour the analysis of scope of (remedial) jurisdiction. This was not done by the lower courts in the *Google Canada* case.

Indeed, in this context, we may seek inspiration, if not guidance, in how states approach the matter of recognition and enforcement of foreign judgments. In traditional Common Law states, for example, recognition at common law is only possible if the foreign court rendering the judgment had jurisdiction on a ground recognised by the relevant jurisdiction, and a foreign court is only regarded as of competent jurisdiction if the defendant was present or resident in the jurisdiction concerned at the time the proceedings were instituted, or in one way or another submitted to the jurisdiction of the foreign court.[41] This is a clear recognition of the link between, on the one hand, the strength of the connection between the forum and the dispute and the parties and, on the other, the willingness to be affected by foreign decisions.

I would go as far as to suggest that we ought to consider moving towards a more mathematical thinking when dealing with this type of situation rather than the more primitive 'tick-box' system courts apply today. Let me explain using a hypothetical scenario.

[37] Reasons for Judgment of the Honourable Justice Fenlon dated 13 June 2014 [104].

[38] ibid [25]. [39] ibid [27]. [40] ibid [28].

[41] Dan Svantesson, *Private International Law and the Internet* (3rd edn, Kluwer Law International 2016) 218–19. A similar reasoning can be found in other states, such as in the US (ibid 257–61).

Imagine that the question before the court is whether or not to order an Internet intermediary to delist certain content globally. Through legal tradition and precedents the court has become accustomed to dealing with such a matter by compartmentalizing the issue into three sub-questions requiring the court to consider (1) whether it has personal jurisdiction over the Internet intermediary in question, (2) whether there are reasons it ought to decline jurisdiction, and (3) whether an issued order legitimately may have global reach. Let us also say that the court concludes that it is just barely convinced that it may claim jurisdiction over the relevant Internet intermediary. On a 0 to 100 scale of how convincing the facts are we may perhaps say 60/100. Using the same type of scale, let us imagine that the court feels that the facts are highly convincing that it need not decline jurisdiction (let us say 90/100). Finally let us imagine that the court recognises that there are quite strong comity reasons suggesting that it ought not grant an order with global effect, but that on balance (60/100 on our scale) it sees it as legitimate to issue such an order.

Under the binary—tick-box—structure the law operates with today we need only consider *whether* the court was convinced on each of the three matters. We do not need to take account of the *degree* to which the facts were convincing, and importantly, we totally ignore the cumulative degree to which the facts were convincing for the overall question of whether or not to order an Internet intermediary to delist the content globally.

If we instead approach the matter as a series of probabilities, the outcome is quite different. We would then perform the following calculation:

$$60/100 \times 90/100 \times 60/100 = 0.324$$

In other words, in a cumulative sense, and seen as mathematical probabilities, the facts of the matter would not support an order requiring the Internet intermediary to globally delist the content.

I hasten to acknowledge that this mathematical approach can be attacked for several reasons, and I am not necessarily advocating it—in this form—as the answer. However, I do think it points to something important—we cannot ignore the fact that the way in which we compartmentalise legal questions may affect the conclusions we reach. A court that simply asks whether on the whole a given set of circumstances makes it legitimate to issue an order with global effect may come to a different conclusion to that of a court that breaks that question down into a set of sub-questions in relation to which it is sufficient that it is just barely convinced by the facts at hand. Compartmentalisation plays an important role in law, and the jurisprudential framework I have advanced for how we approach jurisdiction utilizes compartmentalisation. The point is that we need to be careful how we use compartmentalisation and constantly be willing to re-evaluate its use where the results warrant it.

9.2.2 The Connection between the Party and the Dispute

Even in a situation where all other factors point to a broad scope of (remedial) jurisdiction being legitimate, an order with a broad scope may not necessarily be legitimate against a non-party even where it is (or would have been) legitimate against a party to the dispute. The courts in the *Google Canada* case alluded to the fact that there is a difference between an order against a party to the dispute and an order against a non-party. This ought to have been an important consideration in the *Google Canada* matter as Google was not the target of the litigation per se; rather Google was brought into the matter as a way of seeking to make the original defendant's business uneconomical by making sure it was not found via Google search. However, as already mentioned, this did not save Google on this occasion.

9.2.3 The Impact on Other Countries and Persons in Other Countries

The reality is that with an interconnected world—not least online—it is quite simply impossible to avoid situations where court orders in one country have an effect in other countries. In other words, some 'collateral damage' may be unavoidable.

However, as acknowledged for example by the Court of Appeal, any order that impacts the sovereignty of another country must be carefully assessed as to whether it is nevertheless appropriate: '[C]ourts should be very cautious in making orders that might place limits on expression in another country. Where there is a realistic possibility that an order with extraterritorial effect may offend another state's core values, the order should not be made.'[42] Thus, as already highlighted in the discussion of how we categorise different types of jurisdiction, the greater the impact on foreign countries, and persons in foreign countries, the stronger the reason to limit the scope of (remedial) jurisdiction. This is particularly so where the impact relates to (1) strangers to the lawsuit and/or (2) the fundamental human rights, such as privacy and freedom of expression, of the persons in foreign countries.

[42] Decision of the Court of Appeal of British Columbia dated 11 June 2015 [92].

In this context, it is worth observing that Fenlon J made the following statement supporting the inclusion of this factor in the scope of (remedial) jurisdiction analysis: 'The impact of an injunction on strangers to the suit or the order itself is a valid consideration in deciding whether to exercise the Court's jurisdiction to grant an injunction. It does not, however, affect the Court's authority to make such an order.'[43] However, regrettably she did then not return to this matter in her assessment of whether the order in question should be granted.

Turning to the majority judgment in the Supreme Court of Canada, the following statement is of interest:

> Google's argument that a global injunction violates international comity because it is possible that the order could not have been obtained in a foreign jurisdiction, or that to comply with it would result in Google violating the laws of that jurisdiction is, with respect, theoretical. [. . .] And while it is always important to pay respectful attention to freedom of expression concerns, particularly when dealing with the core values of another country, I do not see freedom of expression issues being engaged in any way that tips the balance of convenience towards Google in this case.[44]

It is interesting to compare this nonchalant attitude to the ongoing discussion of when law enforcement agencies in one state may seek access to data held in another state. As alluded to in Chapter 8, in the latter setting one often hears objections raised that such an access to data amounts to an unlawful interference with the sovereignty of the state where the data happens to be located. However, in the *Equustek* case, the Supreme Court of Canada made an order requiring an alteration to the data held in another country. It seems beyond intelligent dispute that this is much more invasive than is mere access to data.

9.2.4 The Type and Effectiveness of the Order Compared to Alternatives

The appropriateness of granting an order with a broad scope of (remedial) jurisdiction is affected by factors such as the cost of complying with that order, whether the order is limited in time, the availability of less onerous alternative measures and the effectiveness of the order (both in an absolute sense, and as compared to alternative measures).[45]

[43] Reasons for Judgment of the Honourable Justice Fenlon dated 13 June 2014 [147].

[44] *Google Inc v Equustek Solutions Inc* 2017 SCC 34, paras 44–45.

[45] For this factor, I draw upon Justice Arnold's reasoning in *Cartier International AG et al v British Sky Broadcasting Ltd et al* [2014] EWHC 3354 (Ch).

Regarding the available alternative measures in the *Google Canada* case, the most obvious option is, of course, the already established system for blocking that Google Inc provides and that has resulted in hundreds of webpages relating to the networking device in dispute in that matter having been blocked. In this context, it is to be noted that the practice of blocking individual webpages is obviously likely to cause less 'collateral damage' than will the blocking of entire websites as is requested by the plaintiff in that matter.

As to the effectiveness of the order sought by the plaintiff in the *Google Canada* matter, I will restrict my comments to the aspects of relevance for the matter of scope of (remedial) jurisdiction. Extending the order globally was justified by (1) the potential that Canadians circumvent any measures implemented to block content in Canada (eg by visiting non-Canadian Google sites or, where geo-location technologies were applied, by using circumvention tools)[46] and (2) the fact that the majority of sales of the disputed networking device allegedly are to non-Canadian buyers.[47]

As to the first of these, it should be noted that according to the Affidavit of Keller referred to in the Factum of the Appellant, 'In Canada, approximately 95% of Google searches by users are run via www.google.ca, Google's search engine for Canada.'[48] Thus blocking on 'www.google.ca' only would be highly effective on the Canadian market,[49] apart from in relation to those Canadians who are (1) already aware of the disputed networking device and (2) so determined to purchase it so as to change country-version of Google's search engine to search for a website that sells that device.

Extending the blocking globally on Google seems unlikely to deter this latter category of determined buyers as they can change search engine more easily than they can change country-version of Google search.[50] Thus, it

[46] See eg Decision of the Court of Appeal of British Columbia dated 11 June 2015 [21].

[47] See eg ibid [25].

[48] Factum of the Appellant, Google Inc dated 13 June 2016, at n 4.

[49] It is surprising that the lower courts were not satisfied with a 95% effectiveness level in relation to Google users, when they were willing to accept granting an order against Google only—an order that only covers 70–75% of Canadian search engine users according to the market share statistics discussed by Fenlon J (Reasons for Judgment of the Honourable Justice Fenlon, dated 13 June 2014 [152] and the Court of Appeal (Decision of the Court of Appeal of British Columbia dated 11 June 2015 [19]).

[50] After all, as noted by the Court of Appeal: 'Internet users with Canadian IP addresses—a group comprising almost all users accessing the Internet from within Canada—are, by default, redirected from "google.com" to "google.ca" when they perform searches' (Decision of the Court of Appeal of British Columbia dated 11 June 2015 [21]). In contrast, there is of course no function that redirects users from other search engines to Google's search engine.

makes little sense to extend the scope of (remedial) jurisdiction globally for this purpose.

The second reason for a worldwide order—the fact that the majority of sales of the disputed networking device allegedly are to non-Canadian buyers—deserves even closer scrutiny. Even if this amounts to an efficiency reason for granting a worldwide order, this fact has more important consequences speaking against a worldwide order.

The courts overlooked the seemingly obvious fact that, in the light of the majority of sales being to non-Canadian buyers, the court is asked to order a foreign non-party to police alleged infringements occurring in foreign countries perpetrated by defendants beyond the (effective) reach of the court. In such a situation, the connection to the forum can never be strong.

In my view, even in the unlikely event that it is concluded that the connection to the forum is strong enough to meet the 'substantial connection' requirement so as to place the matter within the territorial competence of the court and to make possible a claim of *in personam* jurisdiction over the Appellant, it cannot be strong enough to justify the court adopting such a broad scope of (remedial) jurisdiction as a worldwide order. Thus, in the end, the worldwide order sought ought not to have been granted. Moreover, as to the availability of alternative actions available to Equustek, the dissenting judgment in the Supreme Court of Canada astutely observed:

> In our view, Equustek has an alternative remedy in law. Datalink has assets in France. Equustek sought a world-wide Mareva injunction to freeze those assets, but the Court of Appeal for British Columbia urged Equustek to pursue a remedy in French courts: 'At present, it appears that the proposed defendants reside in France. ... The information before the Court is that French courts will assume jurisdiction and entertain an application to freeze the assets in that country' (2016 BCCA 190, 88 B.C.L.R. (5th) 168, at para. 24). We see no reason why Equustek cannot do what the Court of Appeal urged it to do. Equustek could also pursue injunctive relief against the ISPs, as was done in Cartier, in order to enforce the December 2012 Order. In addition, Equustek could initiate contempt proceedings in France or in any other jurisdiction with a link to the illegal websites.[51]

9.2.5 The Level of Fault of the Party

Where an order is directed at a party that is at fault in some sense, the legitimacy of a broad scope of (remedial) jurisdiction increases with the degree of

[51] *Google Inc v Equustek Solutions Inc* 2017 SCC 34, para 81.

fault attributable to that party. Where the party at which the order is directed is not at fault—as was the case in the *Google Canada* matter—it is more difficult to justify a broad scope of (remedial) jurisdiction.

9.3 Concluding Remarks Regarding Scope of (Remedial) Jurisdiction

In this chapter, I have sought to draw attention to the issue of scope of (remedial) jurisdiction as a separate issue: a separate issue of growing importance due to technological developments, not least the growing role of Internet intermediaries with global, or near global, impact—and an issue of central importance if we are to make progress in the direction of jurisdictional interoperability. However, some will portray these developments more as progress in business models than as legal concerns:

> Google raises the specter of it being subjected to restrictive orders from courts in all parts of the world, each concerned with its own domestic law. I agree with the chambers judge that it is the world-wide nature of Google's business and not any defect in the law that gives rise to that possibility. As well, however, the threat of multi-jurisdictional control over Google's operations is, in my opinion, overstated.[52]

Given the careful attention provided to the scope of (remedial) jurisdiction in the *Google Canada* case, this is a surprising, and disappointingly simplistic, statement by Justice Groberman. Surely the excessively broad scope of jurisdiction claims are legal concerns, or indeed, defects in the law, rather than merely a result of certain adopted business structures? After all, those business structures do not, in any sense, necessitate overly broad scope of jurisdiction claims.

Fortunately, Justice Groberman goes on to stress that: '[c]ourts must, in exercising their powers, consider many factors other than territorial competence and the existence of *in personam* jurisdiction over the parties. Courts must exercise considerable restraint in granting remedies that have international ramifications.'[53] And this brings us to the core of what has been argued here—in addition to appropriate limitations on territorial competence and the existence of *in personam* jurisdiction over the parties, jurisdictional

[52] *Equustek Solutions Inc v Google Inc* [2015] BCCA 265, [69] <http://canlii.ca/t/gjgwv>, accessed 2 April 2017 [56].

[53] ibid [56].

interoperability depends, as a separate issue, on how we delineate scope of (remedial) jurisdiction.

Finally, one important point I have tried to hint at in this chapter is that it is not a matter of global blocking versus no global blocking. The correct approach will always be context-dependent. For example, child abuse materials should be removed globally; and it is my impression that at least the major internet intermediaries are on-board with that. However, global blocking cannot be the default response to all content that happens to violate the law of some state.

10

A Layered Approach to Jurisdiction

A useful step that can be taken towards increased jurisdictional interoperability is for lawmakers to make no broader claims of jurisdiction than what is required. Implementing appropriate limits as to scope of jurisdiction—as discussed in the previous chapter—is only one aspect of this. Here, I describe and discuss another practical measure that may be advantageous in some settings.

It seems to me that lawmakers have a tendency to approach the question of jurisdiction in an overly simplistic, indeed naïve, manner. This is well illustrated in the European Union's General Data Protection Regulation.[1] Article 3 of that Regulation delineates the Regulation's geographical scope of application. And, as commonly is the case for such instruments, it does so in an all-or-nothing manner; that is, where a party's conduct meets the thresholds outlined in Article 3, the (unfortunate) party in question is bound to abide by the entire Regulation with all its diverse provisions. Thus, for example, where a controller not established in the Union processes personal data of data subjects residing in the Union, and the processing activities are related to the offering of goods or services (irrespective of whether a payment by the data subject is required) to such data subjects in the Union, the controller is bound by all the provisions of the Regulation. Similarly, where such a controller processes personal data of data subjects residing in the Union, and the processing activities are related to the monitoring of their behaviour as far as their behaviour takes place within the European Union, the controller has to abide by the Regulation in its entirety.

As data privacy laws always incorporate a diverse range of legal rules—something that is particularly palpable in the General Data Protection Regulation—we ought to dissect each such piece of legislation and identify the different types of rule they include. For example, most data privacy laws contain provisions seeking to discourage, or even penalise, unauthorised and

[1] Regulation (EU) 2016/679 of the European Parliament and of the Council of 27 April 2016 on the protection of natural persons with regard to the processing of personal data and on the free movement of such data, and repealing Directive 95/46/EC (General Data Protection Regulation), [2016] OJ L 119/1–88.

unreasonable disclosure or other use of personal data. Such provisions are similar in nature to rules found in other areas of law, such as the law of defamation. In a sense, they are private in nature and, therefore, fundamentally different to some other types of rules found in some data privacy laws, such as requirements of a designated Data Protection Officer,[2] that are of a public law nature.

In the light of this, it is simply not productive of good results to view the question of the law's jurisdictional reach from a singular perspective. Indeed, it is not sufficient to ask when jurisdictional claims are justified in relation to the relevant law. We need to take one step further and introduce a greater level of nuances and sophistication—we need a *layered approach* taking account of the multifaceted nature of substantive law; in this example, data privacy law.

For example, while it may be reasonable to ask a foreign company to abide by a state's abuse-prevention rules (such as rules discouraging or penalising unauthorised and unreasonable disclosure or other use of personal data) based on a certain degree of contact between that company and that state (eg a transaction involving the collection of personal data), that same degree of contact may not justify that state imposing on the company the duty of designating a Data Protection Officer. Thus, the key to balance and reasonableness in the field of 'extraterritoriality' in data privacy law lies in matching the various provisions found in each data privacy law to suitable criteria for their reach of application.

As often is the case with complex tasks, this particular task may be carried out in more than one manner. Here, continuing to use data privacy law as an illustration, I will perform the task in two steps. First, I will discuss what types of substantive data privacy rules ought to be fitted within each layer of jurisdictional claims. Second, I will identify suitable threshold tests limiting the jurisdictional reach of each of these different layers.

10.1 The Layers

In an attempt to strike a balance between usability and sophistication, I have opted for a model with three different layers of jurisdictional claims. However, I acknowledge that, despite its prevalence in Western fairy-tale tradition, there is no magic in the number three and that one could picture a model with any number of layers from two and up.

2 See eg ibid Arts 37–39.

In any case, for my model here I have chosen to refer to my three layers as:

1. 'the abuse-prevention layer';
2. 'the rights layer'; and
3. 'the administrative layer'.

I will not seek to dissect every single data privacy law and identify which provisions fit in which layer. After all, data privacy law is merely used as an example. In any case, I hasten to acknowledge that such an exercise is by no means free from subjectivity. However, a few illustrative examples can usefully be presented.

As already noted, data privacy laws commonly contain provisions seeking to discourage or to penalise unauthorised and unreasonable disclosure or other use of personal data, in a manner similar to how, for example, defamation law seeks to prevent abuse. Such rules ought clearly to fall within the abuse-prevention layer. Within the rights layer we can comfortably place data privacy rules such as the right of access and the right of correction commonly found in data privacy laws. As also already hinted at, in the administrative layer, we should place data privacy rules such as the requirements of a designated Data Protection Officer[3] found in the proposed EU General Data Protection Regulation.

On a practical level, perhaps the greatest hurdle to overcome in organising the substantive data privacy rules into these three layers is found in the fact that certain such substantive rules, and even more so certain Articles of substantive rules, incorporate matters that may fall within different layers. I do not strive to address that concern here but note that the solution ought to be found in the structuring of the substantive rules and Articles as such. In other words, it may be that a reorganisation of the substantive rules and Articles is required in order that they may be more easily fitted within the layers. This will no doubt be an onerous task requiring great sensitivity and insight.

On a theoretical level, the greatest hurdle to overcome in dividing the substantive data privacy rules into these three layers is found in the fact that we—as far as data privacy is concerned—are dealing with a fundamental human right. Put differently, we may question the suitability, or indeed possibility, of splitting a fundamental human right into different layers of obligations. However, I think we may reasonably view the standard set by international human rights law as a minimum standard, and I suspect we may see initiatives such as the General Data Protection Regulation as partly going beyond that minimum standard. Thus, as long as

[3] See eg ibid Arts 37–39.

the layering does not result in any instances of violations of that minimum standard, no problems should arise.

10.2 Setting Thresholds for the Layers

Having outlined, albeit in a somewhat eclectic manner, the types of substantive data privacy law rules that ought to fit in each layer, these three layers are now discussed in order and proposals are made for how we can delineate their respective jurisdictional reach.

Owing to the similarity to, for example, defamation law, data privacy law provisions falling within the abuse-prevention layer may reasonably be given an equally broad reach of application as defamation law is typically given. Without engaging in a full-scale analysis of the reach of defamation laws, it is undeniable that defamation law is commonly applied in a wide jurisdictional manner.[4] At the same time, even a cursory examination of cross-border Internet defamation cases makes clear that the search for a broadly accepted approach is still on.[5] Consequently, we are here not restricted to adopting jurisdictional delineation applied in defamation disputes.

I argue that, for the abuse-prevention layer, we may apply an unrestricted version of the doctrine of 'market sovereignty' that I have presented in Chapter 6.[6] Thus, a state has market sovereignty, and therefore justifiable jurisdiction, over Internet conduct where it can effectively exercise 'market destroying measures' over the market that the conduct relates to. A state ought to be free to apply the parts of its data privacy law that fall within the abuse-prevention layer to a foreign party as soon as it has market sovereignty over the market to which that party's conduct relates. Essentially, this means that 'extraterritorial' jurisdiction is possible as soon as a foreign controller collects data from an EU resident present in the EU, and one could imagine alternatives to the proposed reliance on the doctrine of market sovereignty that would produce a substantially similar result.

When we turn to the rights layer—layer two—it is clear that we can no longer rely on binary tests, such as the test applied for the abuse-prevention

[4] See eg the High Court of Australia case of *Dow Jones & Company Inc v Gutnick* (2002) 210 CLR 575 and the ECJ joined Cases C-509/09 *eDate Advertising GmbH v X* and C-161/10 *Olivier Martinez and Robert Martinez v MGN Limited*.

[5] See eg Dan Svantesson, *Private International Law and the Internet* (3rd edn, Kluwer Law International 2016).

[6] See also Dan Svantesson, *Extraterritoriality of Data Privacy Law* (Ex Tuto 2013).

layer. We need a test that takes account of the degree of contact between the object (eg a foreign company) and the state seeking to exercise jurisdiction.

Here it is useful to pause to consider the inadequacy of the grounds for jurisdiction typically put forward in international law, and most prominently through the well-known 'Harvard Draft'.[7] As noted in the discussions above, none of the grounds for jurisdiction outlined in the 'Harvard Draft' take account of the degree of contact. Instead they focus on binary measures such as whether the offender[8] or victim was within the territory of the country claiming jurisdiction or not[9] and on whether the offender is a national of that country or not.[10] Such rules are, due to their lack of sophistication, doomed to produce unacceptable outcomes when applied to complex scenarios. Thus, the answer to the question of what test we may apply to the rights layer is not to be found in the grounds for jurisdiction typically put forward in international law.

To find an appropriate test for layer two claims of extraterritorial jurisdiction—the rights layer—we may instead turn for example to basic US law principles of long-arm jurisdiction and, more specifically, to the familiar 'minimum contact' test first formulated in the *International Shoe Co v Washington*:[11] '[D]ue process requires only that in order to subject a defendant to a judgment in personam, if he be not present within the territory of the forum, he have certain *minimum contacts* with it such that the maintenance of the suit does not offend '*traditional notions of fair play and substantial justice*".'[12] The reason for the, at that time, new jurisdictional rules stated in *International Shoe Co v Washington* was the aim of an adjustment to a more mobile society—an aim of even greater importance today with our significantly more mobile society.

Since the case was decided on 3 December 1945, the factors taken into account in the test have changed somewhat over time, and different courts have interpreted the *International Shoe Co v Washington* case differently. That does not, however, impact its use here as I am not proposing an adoption of the exact test as such. Rather, what I am proposing is that the idea of a minimum contact test is also useful for delineating the jurisdictional reach of the substantive data privacy rules that fall within the rights layer. Having said this, important guidance can be gained from one of the most important cases

7 'Introductory Comment to the Harvard Draft Convention on Jurisdiction with Respect to Crime 1935' (1935) 29 Supp AJIL 443.

8 The subjective territoriality principle.

9 The objective territoriality principle.

10 The nationality principle.

11 *International Shoe Co v Washington* 326 US 310 (1945).

12 ibid 316 (emphasis added).

examining the minimum contact test—*Hanson v Denckla*.[13] In that case, the Court states that it is always necessary that there be an act 'by which the defendant purposefully avails itself of the privilege of conducting activities within the forum state, thus invoking the benefits and protections of its laws',[14] for a minimum contact to be established.[15] This reasoning may usefully be adopted also for the minimum test for layer two.

Turning to the search for a suitable test for the administrative layer, we again need a test that is more sophisticated than the binary tests of the 'Harvard Draft', and we may again draw upon ideas and concepts found in basic US law principles about jurisdiction. More specifically, to account properly for the onerous nature of the substantive data privacy rules that fall within the administrative layer—layer three—we must ensure that the test we apply is strict indeed, and we can benefit from looking to how the United States has dealt with so-called 'general jurisdiction' of courts. Under US law, general jurisdiction may be exercised when a party's ties to the forum are continuous, systematic, and ongoing.[16] This is measured by reference to the quality and quantity of the defendant's contacts with the forum.[17] As noted for example by Jensen, this means that a fairly high standard must be met: 'To assert general jurisdiction over an out-of-forum defendant, a court must find that the defendant's minimum contacts with the forum are so systematic that they serve as a proxy for physical presence.'[18] Furthermore, '[e]ven if substantial, or continuous and systematic, contacts exist, the assertion of general jurisdiction must be reasonable.'[19]

There are a limited number of US cases relating to the assertion of general jurisdiction in an international context, one of the most well-known being *Helicoptoros Nacionales De Columbia SA v Hall*.[20] In that case, the Supreme Court of the United States denied a Texas Court the right to assert jurisdiction over a non-resident defendant. The contact between the state of Texas and the defendant included the following:

- the contract pursuant to which the defendant provided the transportation service being used at the time of the crash was negotiated in Houston, Texas;

[13] *Hanson v Denckla*, 357 US 235 (1958). [14] ibid 253. [15] ibid.

[16] *Helicoptoros Nacionales De Columbia, SA v Hall*, 466 US (1984).

[17] *Phillips Exeter Acad v Howard Phillips Fund, Inc*, 196 F3d 284, 288 (1st Cir 1999).

[18] JM Jensen, 'Personal Jurisdiction in Federal Courts over International E-commerce Cases' (2006–2007) 40 Loy LAL Rev 1507, 1521.

[19] *Gator.com Corp v LL Bean*, 03 CDOS 7986, 2003 US App LEXIS 18115 (9th Cir 2 September 2003) referring to *Amoco Egypt Oil Co v Leonis Navigation Co, Inc*, 1 F3d 848, 852–53 (9th Cir 1993).

[20] *Helicoptoros Nacionales De Columbia, SA v Hall*, 466 US (1984).

- eighty per cent, or US$4,000,000-worth, of the defendant's fleet of helicopters were purchased from Bell Helicopters Company in Fort Worth, Texas; and
- the defendant sent its prospective pilots, management, and maintenance personnel to Fort Worth, Texas, for training purposes.

Although it could be argued that these facts indicate an ongoing continuous and systematic contact with the state of Texas, the Court held against the assertion of general jurisdiction. This was motivated by a range of factors, the most noteworthy being that 'mere purchases, even if occurring at regular intervals, are not enough to warrant a State's assertion of in personam jurisdiction over a non-resident corporation in a cause of action not related to those purchase transactions.'[21]

Since there was some form of ongoing continuous and systematic contact with the forum state in this case, one must ask why the assertion of general jurisdiction was not upheld. While it is possible that the 'reasonableness' influenced the Court, this case illustrates that ongoing, continuous, and systematic contacts must fulfil some form of *substantiality requisite*—there must be a certain degree of substantiality to the ongoing continuous and systematic contact between the forum and the case at hand.

To conclude this part of the discussion, I propose that the test for the reach of application of the substantive data privacy rules that fall within the administrative layer should be focused on whether the object's contacts with the country claiming jurisdiction are sufficiently substantial, continuous, and systematic, making the exercise of extraterritorial jurisdiction reasonable. This proposal does not, however, amount to an adoption of the US concept of general jurisdiction as such. It merely draws upon the test developed for the purpose of delineating US general jurisdiction.

Importantly, none of the tests in the three layers are comparative; that is, there is never any need to consider whether a particular matter is more closely connected to another forum. Thus, all three tests are distinctly unilateral in nature, which ought to simplify their application in practice.

10.3 A Model Article

At this stage it may be useful to try to summarise what has been said already about the layered approach advocated here, in the form of an Article that could form part of data privacy legislation around the world:

[21] ibid.

Article X

1. The following Articles of this [Regulation] [22] fall within layer one (the 'abuse-prevention layer'): [list relevant Articles].
2. The following Articles of this [Regulation] fall within layer two (the 'rights layer'): [list relevant Articles].
3. The following Articles of this [Regulation] fall within layer three (the 'administrative layer'): [list relevant Articles].
4. The Articles of this [Regulation] falling within layer one apply in an extraterritorial manner where [the Union] has market sovereignty over the [controller's] conduct, by reference to whether it can effectively exercise 'market destroying measures' over the market that the conduct relates to.
5. The Articles of this [Regulation] falling within layers one and two apply in an extraterritorial manner where a [controller] not established [in the Union] has certain minimum contacts with [the Union] for example by purposefully availing itself of the privilege of conducting activities within [the Union], thus invoking the benefits and protections of its laws.
6. The Articles of this [Regulation] falling within layers one, two, and three apply in an extraterritorial manner where a [controller's] ties to [the Union] are sufficiently substantial, continuous, and systematic to make the exercise of extraterritorial jurisdiction reasonable.

To this could be added a provision such as this:

7. This [Regulation] does not apply in an extraterritorial manner to a [controller] that, acting in good faith, has taken reasonable steps to avoid contact with [the Union].

This is obviously only one possible way in which one may structure the type of layered approach I have canvassed. And admittedly, some aspects of this Article X could benefit from being fleshed out somewhat. Nevertheless, it is hoped that 'Article X' constitutes a useful illustration of the layered approach.

A few examples may usefully illustrate how the layered approach advocated would work on a practical level. Imagine that company Z, based in the United States, enters into thousands of transactions with consumers in Europe every month and directs marketing campaigns towards European consumers online, in European magazines, and on European TV stations. Imagine further that it undertakes no marketing activities towards Australia and comes into contact with only one Australian consumer who contacts the company to gain information about one of the company's products. Finally, imagine that company Z regularly enters into a small number of transactions

[22] Apart from the reference to 'list relevant Articles', the text in brackets is aimed directly at the EU context and should obviously be substituted where Article X is applied in other jurisdictions.

with consumers in Singapore but does not in any sense target its activities towards Singaporean consumers.

In such a situation, it could be concluded that, under the layered approach outlined here, company Z would fall within all three layers in relation to EU data privacy law. Further, company Z would fall within the first two layers in relation to Singaporean data privacy law and only fall within the first layer—the abuse-prevention layer—in relation to Australian data privacy law.

While the model of a layered approach to extraterritoriality in data privacy law outlined ought to have obvious benefits, it is also associated with at least one problem as can be illustrated in the following example.

Imagine that person A and person B, both residing in the EU, are considering buying a particular eBook. Person A does so from company X, and person B does so from company Y. Imagine further that, while company Y deals with EU consumers to such a degree that it meets the minimum contact test of the rights layer, company X does not meet that test.

In this case, person A is then enjoying a lower level of protection (ie only abuse-prevention), while person B is enjoying a higher level of protection (ie both abuse-prevention and the rights of the rights layer). This difference in protection may be seen as unfair since, from the perspective of persons A and B, the level of protection they respectively receive must appear rather random; that is, persons A and B may not, at the time of interaction with companies X and Y respectively have known the extent to which those companies engage on the relevant market. This disadvantage does not, however, substantially undermine the overall attractiveness of the model put forward here.

10.4 Concluding Remarks Regarding the Layered Approach

This chapter observes how it may be inappropriate to apply a single jurisdictional threshold to diverse instruments such as, for example, data privacy laws. In the light of this observation, a proposal was outlined for a 'layered approach', under which the substantive law rules of such instruments are broken up into different layers, and how different jurisdictional thresholds may then be applied to each such layer.

While the layered approach I advocate has been discussed here in the data privacy context, it must be clear that it offers the option of a valuable sophistication for any complex and diverse law. Nevertheless, it is perhaps particularly useful for any legislation, or other law, that covers diverse matters.

Before concluding this discussion, one additional observation must be made. While the layered approach is here discussed primarily as a technique to be utilized in legal drafting, it may also be applied in the interpretation and application of legal rules. For example, remaining with the EU's GDPR it is still possible to adopt the layered approach for the interpretation and application of Article 3 even though that approach was not adopted in the drafting of Article 3. Indeed, it may be argued that the general proportionality thinking characteristic of EU law mandates the adoption of the layered approach for the interpretation and application of Article 3.

Regardless of the stage at which the approach is utilized, applied appropriately, the layered approach will be a valuable tool towards jurisdictional interoperability.

11

The Role of Geo-location Technologies

My interest in geo-location technologies dates back to 2002, and I published my first article on the topic in 2004 in the *John Marshall Journal of Computer & Information Law*.[1] In fact, as far as I have been able to ascertain, that was the first law journal article devoted entirely to the topic of the legal implications of geo-location technologies, at least in English. Since then I have devoted much of my research efforts to the intersection, and interaction, between law and geo-location technologies. For a long time, I was astonished by the legal community's resistance to taking geo-location technologies seriously. For example, there are numerous examples of court decisions that ought to have considered, or at least taken account of, geo-location technologies, but that for one reason or another do not do so. Consider, for example, the early Australian example of *Macquarie Bank Limited & Anor v Berg*.[2] There the Supreme Court of New South Wales noted that:

> [t]he limitation [to publication occurring in New South Wales only] is ineffective. Senior counsel [for the plaintiffs] acknowledged that he was aware of no means by which material, once published on the Internet, could be excluded from transmission to or receipt in any geographical area. Once published on the Internet material can be received anywhere, and it does not lie within the competence of the publisher to restrict the reach of the publication.[3]

Clearly the outcome could have been quite different if senior counsel for the plaintiffs and the Court had been conscious of the geo-location technologies already in existence at the time of the dispute in question. While the *Macquarie Bank Limited & Anor v Berg* case is a valuable reminder of the importance of lawyers and courts being up to speed with technological developments, one may nevertheless be tempted to put this down as an example of it being unsurprising that at least the Court was unaware of geo-location technologies.

[1] Dan Svantesson, 'Geo-Location Technologies and Other Means of Placing Borders on the "Borderless" Internet' (Fall 2004) XXIII(1) John Marshall J Compu & Info L 101–39.

[2] *Macquarie Bank Limited & Anor v Berg* [1999] NSWSC 526.

[3] ibid [12].

Solving the Internet Jurisdiction Puzzle. First Edition. Dan Jerker B. Svantesson.

In contrast, it is difficult to point to any mitigating factors where the CJEU, in 2010, opted to ignore the highly relevant geo-location technologies in the combined cases of *Pammer* and *Hotel Alpenhof*.[4] This is especially so given that one of the parties had expressly argued: 'Account should be taken of the characteristics of the internet, which makes it impossible to restrict information to Austrian territory;'[5] and that the Court in the same year emphasised that measures to block access to an 'internet site for persons residing in [a specific Member State] […], is an indispensable element of the protection in respect of games of chance that is intended to be provided by the [Member State] within its own territory'.[6] Remarkably, this crucial issue was not discussed further at all, neither by Advocate General Trstenjak nor by the Court in *Pammer* and *Hotel Alpenhof*. There can, of course, be no doubt about the relevance a use of geo-location technologies, or lack thereof, must have for any proper assessment of 'targeting' as was the central matter in the case.[7]

At any rate, it is interesting to observe how we now have gone from a situation where the use and utility of geo-location technologies commonly was ignored, or argued to be irrelevant (typically due to an underappreciation of their accuracy), to a situation where the same technologies are taken for granted—there was, as far as I can discern, no expression of any gradual awakening to these technologies.

Today, there is consequently little point in arguing *that* the law needs to take account of geo-location technologies; the debate has moved on, and this is now nothing more than a widely recognised truism.[8] Here, I will seek

[4] Joined cases C-585/08 and C-144/09.

[5] Opinion of Advocate General Trstenjak delivered on 18 May 2010, para 32.

[6] C-258/08, para 43.

[7] See, however, Proposal for a Regulation of the European Parliament and of the Council on addressing geo-blocking and other forms of discrimination based on customers' nationality, place of residence or place of establishment within the internal market and amending Regulation (EC) No 2006/2004 and Directive 2009/22/EC, COM(2016) 289 final.

[8] I hasten to acknowledge, however, that geo-location technologies are not taken into account quite to the degree one might have assumed that they would be. Consider, for example, the 'scope of jurisdiction' debate that has occurred in the wake of the *Google Spain*—'right to be forgotten'—decision. The CJEU was silent on the geographical scope of the right to be forgotten, or more appropriately, the right to delisting. Much could be said about the clash between different views on scope of jurisdiction that has occurred, but for our purposes here the most interesting aspect is the fact that the debate has become so focused on top-level domains at the expense of geo-location. For example, Article 29 Data Protection Working Party, 'Guidelines on the Implementation of the Court of Justice of the European Union Judgment on "Google Spain and Inc v Agencia Española de Protección de Datos (AEPD) and Mario Costeja González"—C-131/12' (2014) WP225, 9. This shaped much of the debate; see, however, Brendan Van Alsenoy and Marieke Koekkoek, 'Internet and Jurisdiction after Google Spain: The Extraterritorial Reach of the "Right to be Delisted"' (2015) 5(2) Intl Data Priv L 105.

instead to discuss in more detail *how* the law on jurisdiction needs to relate to geo-location technologies.

11.1 Briefly about what Geo-location Technologies Are, and Are Not

Despite the fact that the legal debate has now generally embraced the importance of geo-location technologies, there is a degree of confusion as to what we are actually dealing with. While geo-location technologies are being used for a diverse range of purposes, such as fraud detection, authentication, content targeting, security, and network efficiency, it is primarily the use of geo-location technologies for conditional access and legal compliance that are of concern here. If providers of Internet content are able to identify the geographical location of their visitors, they can potentially comply with the local regulations of the states from which their visitors come. Indeed, as the content provided can be adjusted depending on the visitors' geographical locations, geo-location has the advantage of providing content providers with the means to comply with multiple, varying, and even contradictory local regulations. This may, obviously, go some way towards the goal of jurisdictional interoperability, and the way in which the law relates to geo-location technologies may either harness this tool towards the goal of jurisdictional interoperability or, indeed, discourage its use.

It is interesting to consider here the actual uptake of geo-location technologies by providers of Internet content. One only has to surf the net for a short period to note that a large amount of content appears to be provided without any geographical restrictions. It may still be the case that only a comparatively small number of content providers apply geo-location technologies.

However, when some of the users of geo-location technologies include the largest Internet companies in the world (eg Google), the practical effect is that most Internet users come into daily contact with geo-location technologies. In other words, even if the number of content providers using geo-location technologies may be relatively small, a high percentage of Internet users come into daily contact with them. And the fact that we mainly access content via search engines—that use geo-location technologies—means that even though most Internet content is *provided* in a location-neutral manner, it is nevertheless *accessed* in a location-dependent manner. Furthermore, all types of apps are showing an immense thirst for ascertaining our geographical locations, and as we move towards an increasingly app-dominated

Internet use, the extent to which we are exposed to geo-location technologies will inevitably increase.

Finally, we must distinguish between geo-location technologies on the one hand and, on the other, the type of filtering some states engage in placing virtual borders around 'their' respective parts of the network—the 'Great Firewall of China' being the most famous example. Such state-run filtering clearly represents a possible tool for jurisdictional interoperability, and the argument can be made that states must be seen to have accepted all the content they do not block out. Similarly, situations where courts order local Internet service providers to block access to certain sites for their users raise different types of question in situations where, for example, youtube.com is ordered to make inaccessible certain specific content. Also these types of action may be important ingredients in achieving jurisdictional interoperability. However, such topics will not be pursued further here. Rather, the geo-location technologies we discuss here are those implemented by the content or service provider, typically by private actors.

11.1.1 Terminology

Geo-location technologies operate on several levels. Elsewhere,[9] I have suggested that it is necessary to distinguish between sophisticated and unsophisticated geo-location technologies. I have also argued that it is useful to distinguish between two categories of sophisticated geo-location technologies: server-side geo-location and client-side geo-location. More recently, I have also pointed to the difference between, on the one hand, 'strict geo-location blocking' and, on the other, a 'country lens approach'.[10] Most importantly, however, and all I want to emphasis here, is that we must always remember that geo-location technologies come in many different forms, ranging from examining the browser language settings to the translation of IP addresses into geographical locations, from reliance on Global Positioning System (GPS) chips and the triangulation of nearby network towers to the mapping of Wi-Fi hotspots.

Recent attention directed at geo-location technologies has displayed a preference for the term 'geo-blocking'.[11] It should, however, be recognised

[9] See eg Dan Svantesson, *Private International Law and the Internet* (3rd edn, Kluwer Law International 2016); Dan Svantesson, 'Time for the Law to take Internet Geo-location Technologies Seriously' (December 2012) 8(3) J Priv Intl L 473–87.

[10] Dan Svantesson, 'Delineating the Reach of Internet Intermediaries' Content Blocking—"ccTLD Blocking", "Strict Geo-location Blocking", or a "Country Lens Approach"?' (2014) 11(2) SCRIPT-ed 153–70, 164–68.

[11] See eg Proposal for a Regulation of the European Parliament and of the Council on addressing geo-blocking and other forms of discrimination based on customers' nationality, place of

that geo-blocking merely represents one way in which geo-location technologies may be applied.

11.1.2 Accuracy

The accuracy of sophisticated geo-location technologies is difficult to gauge and has been a topic of dispute for a long time.[12] However, already in the French Court's judgment in the 2000 *Yahoo!* case,[13] experts concluded that 'it may be estimated in practice that over 70% of the IP addresses of surfers residing in French territory can be identified as being French.'[14] In the context of accuracy, a few key points need to be observed.

First, the accuracy of geo-location technologies is affected by two distinct factors: 'source accuracy' and 'circumvention'. The former relates to the problems associated with building up and/or collecting accurate geo-location data. The 'garbage-in, garbage-out' doctrine has clear application. The latter relates to the fact that it will presumably always be possible to circumvent geo-location technologies. And while some circumvention techniques are technologically advanced, others are easy enough to be used by virtually anyone. The reality is that today geo-location technologies are typically built on highly reliable location data, and the only major accuracy issue stems from the potential for circumvention.

Yet, when discussing the role of circumvention, it must be noted that there are different degrees of incentive for circumventing geo-location technologies.

residence, or place of establishment within the internal market and amending Regulation (EC) No 2006/2004 and Directive 2009/22/EC, COM(2016) 289 final; Ramon Lobato and James Meese (eds), *Geoblocking and Global Video Culture* (Institute of Network Cultures 2016).

[12] See further Dan Svantesson, 'How Does the Accuracy of Geo-location Technologies Affect the Law?' (2008) 2(1) Masaryk UJL & Tech 11–21; Marketa Trimble, 'The Future of Cybertravel: Legal Implications of the Evasion of Geolocation' (2012) 22 Fordham Int Prop, Media & Ent LJ 567; *Nitke v Ashcroft*, 253 F Supp 2d 587 (SDNY 24 March 2003) (NO 01 CIV 11476 (RMB), expert testimony by Seth Finkelstein (10 November 2003) <http://sethf.com/nitke/ashcroft.php>, accessed 4 April 2017; *ACLU v Gonzales*, 478 F Supp 2d 775 (ED Pa 2007) [182]–[6]; Y Shavitt and N Zilberman, 'A Study of Geolocation Databases' (2010) <http://arxiv.org/abs/1005.5674>, accessed 4 April 2017, 3; B Edelman, 'Shortcomings and Challenges in the Restriction of Internet Retransmissions of Over-the-Air Television Content to Canadian Internet Users' http://cyber.law.harvard.edu/people/edelman/pubs/jump-091701.pdf, accessed 4 April 2017, 6; JA Muir and PC van Oorschot, *Internet Geolocation and Evasion* (10 April 2006) <http://www.ccsl.carleton.ca/~jamuir/papers/TR-06-05.pdf>, accessed 4 April 2017; KF King, 'Personal Jurisdiction, Internet Commerce, and Privacy: The Pervasive Legal Consequences of Modern Geolocation Technologies' (2011) 21 Albany LJ Sci & Tech 61, 71.

[13] *International League Against Racism & Anti-Semitism (LICRA) and the Union of French Jewish Students (UEJF) v Yahoo! Inc,* County Court of Paris, interim court order of 20 November 2000.

[14] ibid. However, it would seem that one of the experts, Ben Laurie, later felt a need to explain his statement (B Laurie, *An Expert's Apology* (on file with author)).

The risk of circumvention corresponds directly to the perceived value of the content that can be accessed through the circumvention. Therefore, for example, websites containing popular but geographically restricted copyright-protected content such as movies or music, only released in some countries but not others, are obvious targets for circumvention. But in the context of other uses of geo-location technologies—such as their potential use to geographically separate the use of one top-level domain name[15]—the incentive for circumvention is missing entirely.

Second, when assessing the accuracy of methods for geo-identification, it is important to avoid placing the focus on the marketing-driven *average* accuracy rates presented by the companies behind the method in question and instead pay attention to the *context-specific* accuracy rate. Placing the focus on these will inevitably complicate and increase the cost of court proceedings in that expert evidence may be required in each individual case. However, the importance of ensuring that the courts base their decisions on the accuracy rate that is relevant in the particular case at hand cannot be overstated.

Third, accuracy rates are time-specific as there is a need for keeping geo-location databases up to date. This also emphasises the need for courts to view each matter on a case-by-case basis and not be led astray by average accuracy levels.

In the light of this discussion, while it will presumably always be possible to circumvent geo-location technologies, these technologies do not need to be 100 per cent accurate for most uses, and it consequently does not always matter that they can be circumvented by a limited group of people motivated to do so.[16] The conclusion must be that geo-location technologies are accurate enough that the law must take them into regard, albeit that these technologies may not always be determinative.

11.2 How Geo-location Technologies May Help Us with Internet Jurisdiction

The most obvious—and arguably overly simplistic—way to look at the question of how geo-location technologies may help us with the problems we face

[15] See further Dan Svantesson, 'Is Sharing the Answer to.BRAND Top Level Domain Disputes?' (CircleID, 17 June 2015) <http://www.circleid.com/posts/20150617_is_sharing_the_answer_to_brand_top_level_domain_disputes/>, accessed 4 April 2017.

[16] JL Goldsmith and AO Sykes, 'The Internet and the Dormant Commerce Clause' (2001) 110 Yale LJ 812; see also L Lessig, 'The Zones of Cyberspace' (1996) 48 Stan L Rev 1405 (footnotes omitted).

in the arena of Internet jurisdiction is to focus on the significance of content being available to Internet users at a particular location. If a newspaper is sold in Sydney, we may reasonably assume that the publisher has voluntarily and consciously made the decision to sell its newspaper on the Sydney market and can therefore reasonably be expected to abide by the laws of the Australian state of New South Wales. However, the mere fact that a newspaper website is accessible in Sydney does not allow us to draw such a conclusion. The difference is found in the mode of distribution.

A physical newspaper sets out with a 0 per cent geographical distribution coverage—it reaches no one on its own. And for that percentage to increase, active steps are required of the publisher. In contrast, a website sets out with virtually a 100 per cent geographical distribution coverage in that, once uploaded, it can be accessed globally unless active steps are taken to limit the distribution coverage.

However, the picture painted does not take account of geo-location technologies. Through the use of geo-location technologies we place the publisher of a website in the position to predetermine the desired publications coverage. After all, contact with readers at any undesired locations can be readily avoided via the use of such technologies. And in the light of this, we can arguably apply the same presumption as we do in relation to offline publications; that is, the mere fact that a website is accessible in Sydney may be seen to indicate that the publisher has voluntarily and consciously made the decision to place the website on the Sydney market and can therefore reasonably be expected to abide by the New South Wales laws. Put bluntly, approached carefully, the erection of borders on the Internet through technological means has the potential of legitimising some of the current wide jurisdictional claims.

11.3 Misguided Fear of Geo-location Technologies

Geo-location technologies may be used for 'good', they may be used for 'evil', and they may be used in an arguably pointless manner that is neither good nor bad. The same is true about a hammer, a banana, and a greetings card. However, this plain fact has not prevented attention being placed on geo-location technologies where they are being used for inappropriate purposes.

For example, the European Commission has drawn attention to the extent that geo-location technologies are used as tools for price differentiation:

> To analyse the extent of price discrimination the Commission initiated a web-scraping study in 2013. The study showed that a leading car rental company consistently used to set different prices based on the consumer's place of residence (the

situation has improved since then). For an identical car rental in a given country, a car could at the time be up to 53% more expensive if the consumer resided in the UK compared to Romania, Slovenia and Poland.[17]

While price differentiation is more clearly a legal issue within the European Union than it is between countries not tied to the European Union, consumers will doubtless wish for limitations on the use of geo-location technologies for price differentiation on a more general level, not least given that the online environment lends itself to cross-border price comparisons. Thus, the (mis) use of geo-location technologies for price differentiation ought, of course, to be prevented. However, to focus on the tool is, arguably, to misconstrue the problem. Yet, the European Commission announced that it 'will make legislative proposals in the first half of 2016 to end unjustified geo-blocking'.[18]

The result is the Proposal for a Regulation of the European Parliament and of the Council on addressing geo-blocking and other forms of discrimination based on customers' nationality, place of residence, or place of establishment within the internal market.[19] This proposed Regulation is specifically aimed 'to give customers better access to goods and services in the Single Market by preventing direct and indirect discrimination by traders artificially segmenting the market based on customers' residence'.[20] This aim is of course admirable. However, there are both important limitations to, and serious concerns about, the proposed Regulation.

The Regulation will not affect situations where a trader sells goods and those goods are delivered cross-border to the Member State of the customer by the trader or on his or her behalf.[21] Similarly, situations where a trader provides electronically supplied services, the main feature of which is the provision of access to and use of copyright protected works or other protected subject matter, are also excluded.[22] While both these exemptions—or at least

[17] Commission Staff Working Document: A Digital Single Market Strategy for Europe—Analysis and Evidence Accompanying the Document Communication from the Commission to the European Parliament, the Council, the European Economic and Social Committee and the Committee of the Regions: A Digital Single Market Strategy for Europe {COM(2015) 192 final} Brussels, 6 May 2015, 22.

[18] Brussels, 6 May 2015 COM(2015) 192 final Communication from The Commission to the European Parliament, the Council, the European Economic and Social Committee and the Committee of the Regions: A Digital Single Market Strategy for Europe {SWD(2015) 100 final}, 6.

[19] Proposal for a Regulation of the European Parliament and of the Council on addressing geo-blocking and other forms of discrimination based on customers' nationality, place of residence or place of establishment within the internal market and amending Regulation (EC) No 2006/2004 and Directive 2009/22/EC, COM(2016) 289 final.

[20] ibid 2. [21] ibid Art 4(1)(a). [22] ibid Art 4(1)(b).

the first—may be sensible, they nevertheless impose significant limitations on the usefulness and impact of the Regulation.

More importantly, despite the European Commission's express assurance that '[t]he proposal is compatible with existing Union law on applicable law and jurisdiction',[23] there can be no doubt that there is an obvious tension between, on the one hand, this proposed Regulation, and on the other, the Brussels I bis Regulation and the Rome I Regulation.[24] Put simply, the former—the proposed geo-blocking Regulation—demands that a business should not take steps to avoid contact with consumers based on their location, while the latter say that if a business has directed its activities to the consumer's Member State, it is bound by the laws of that Member State, must expect to be sued there, and may only sue the consumer in that Member State.

The proposed Regulation makes clear that '[c]ompliance with this Regulation shall not be construed as implying that a trader directs his or her activities to the Member State where the consumer has the habitual residence or domicile.'[25] Nevertheless, only the most devoted fans of fairy tales can possibly imagine that this intersection will be free from grey zones. And not even such escapists could possibly dispute the fact that the proposed Regulation, in removing the use (or lack thereof) of geo-location technologies from the targeting assessment, weakens the already unruly targeting test.

Regrettably, the problems with the proposal do not end there. Building on the EU's established appetite for—in a jurisdictional sense—biting off more than it can chew, the Explanatory Memorandum promulgates that '[t]he territorial scope is designed to equally include traders established in the EU and those established in third countries but selling or seeking to sell goods and services to customers in the Union.'[26] An examination of what this means in practice for the world outside the EU, creates a déjà vu-like experience with the thoughts going to the poor craftsmanship of the territorial reach of the initial proposal for the General Data Protection Regulation.

[23] ibid 3.

[24] Regulation (EU) No 1215/2012 of the European Parliament and of the Council of 12 December 2012 on jurisdiction and the recognition and enforcement of judgments in civil and commercial matters (Brussels I bis) and Regulation (EC) No 593/2008 of the European Parliament and of the Council of 17 June 2008 on the law applicable to contractual obligations (Rome I), respectively.

[25] Proposal for a Regulation of the European Parliament and of the Council on addressing geo-blocking and other forms of discrimination based on customers' nationality, place of residence, or place of establishment within the internal market and amending Regulation (EC) No 2006/2004 and Directive 2009/22/EC, COM(2016) 289 final, Art 1(5).

[26] ibid 7.

Let us say that an Australian online trader has an established market in Ireland and has gone through the expense of getting legal advice so as to be comfortable in that it complies with Irish law. Let us also say that this online trader does not sell to consumers in any other European state. After the introduction of the proposed Regulation, that Australian online trader would no longer be allowed to block consumers from any EU Member State.[27] Indeed, if the trader provides electronically supplied services, a main feature of which is the provision of access to and use of copyright protected works or other protected subject matter, it is forced to accept orders from consumers from *all* EU Member States (unless, of course, it stops selling to consumers in Ireland).[28]

Proponents of the approach taken by the EU may stress the fact that the EU market is a single market and non-EU traders enter it on a take it or leave it, all or nothing, basis. However, the harshness of this approach will not take us closer to jurisdictional interoperability and is bound to spark negative reactions.

11.4 The Real Threat of Geo-location Technologies, and Why We Still Need Them

Already in 2004 I acknowledged that it makes sense for lawyers to advise their clients to implement geo-location technologies. However, I also remarked that, as an academic (and as an Internet user), I have concerns about how we use geo-location technologies: '[I]t seems likely that geo-location technologies will contribute to transforming the Internet as we know it into something that more closely resembles our world, so divided by borders of different kinds.'[29] This is a very real concern. Why would a small Swedish newspaper make its content available worldwide if doing so generates little profit but exposes the newspaper to the defamation laws, censorship laws, and other content laws of all the states in the world?

It seems clear that consequences of a severe nature are in store if the use of geolocation technologies becomes the norm rather than the exception and they are applied thoughtlessly. But while geolocation technologies transforming the Internet as we know (or 'knew') it into something that more closely resembles our 'real' world is far from ideal, it may nevertheless be unavoidable—and indeed, perhaps even the best option.

[27] ibid Art 3(1). [28] ibid Art 4(1)(b). [29] Svantesson (n 1) 137.

It is here we must disentangle two assumptions. While it is clear that the Internet has become increasingly location-dependent in that the type of content one obtains depends on one's geographical location, that does not mean that we must abandon the hope of a (largely) borderless Internet. In other words, we may have to accept that not all content can be available at all locations, but the Internet must continue to operate without unnecessary border controls. The reality that different states have different substantive laws simply cannot be ignored, and the regulation of activities on the Internet must, in one way or another, take account of this reality. And as long as the rules of jurisdiction are focused on the location of the effect rather than on the location of the acting party, there is a significant law-driven incentive for providers of Internet content to know the location of those who access their content. Currently the most effective way of gaining such knowledge is through the application of some form of geo-location technology.

In the many articles I have written on the topic of geo-location technologies, I have consistently been sitting firmly on the fence on the question of the vices and virtues of the use of such technologies. And while sitting on the fence may be a rather lonely experience, it does give you a very good overview. Now, however, I suspect that—despite the impact it will have—the use of geo-location technologies is indeed unavoidable.

I cannot point to any individual statement or event that has convinced me that we have now reached a point where the use of geo-location technologies has become necessary. But having had the privilege of listening to leading academics, industry representatives, policy-makers, and practitioners, I am concerned that, unless we adopt widespread geo-location technologies, the Internet is facing even greater threats than those posed by the use of such technologies.

The key reason for my concern is the extent to which states are requesting global blocking based on infringements of local law; in other words, the tendency of excessively broad claims of scope of jurisdiction. Examples of this were discussed in Chapter 9, but the examples provided are merely the tip of the proverbial iceberg. In other words, the threat is this: unless the Internet is partially fragmented through geo-location, there is a real risk that much Internet content—content that most people in democracies value—will be blocked worldwide based on it being contrary to some law somewhere in the world.

If we accept this conclusion, we must necessarily also be prepared to give up a range of privileges we have become used to having—the Internet would in a sense become much more regionalised, or Balkanised. At the same time, it must be pointed out that examples can be found illustrating that the Internet can be invigorated by the use of geo-location technologies.

Up until Athens 2004, the web-coverage of the Olympic Games had been limited to text and pictures in various forms.[30] While we had been able to find impressive statistical data, biographical details about all athletes, and various other features, there had been a paucity of audio and video coverage. The International Olympic Committee (IOC) provided two reasons for this: the 'poor quality' of Internet broadcasts and the lack of reliable geo-location technologies.[31] It has been reported that the former IOC vice-president[32] stated:

> Until the technology changes to allow the video to be restricted, we have a problem', [then IOC vice-president] Pound said. 'Historically, we have sold rights in a particular territory. Unless and until you can guarantee that the signal will be restricted to your territory, then you cannot put real time video or real time audio on the Internet.[33]

This changed once geo-location technologies were regarded as sufficiently accurate for the IOC's purposes. In other words, the use of geo-location technologies enabled certain content to be made available online.

11.5 Will Courts Allow Geo-location Technologies?

Whether geo-location technologies will save us from excessive worldwide blocking orders remains to be seen. To a degree, this matter may depend on how well the courts manage to understand the mathematics involved, and unfortunately lawyers are notoriously bad at mathematics.

In the mentioned *Google Canada* case, for example, the lower court displayed a remarkable mathematical inconsistency: a mathematical inconsistency that may affect how geo-location technologies are viewed. Based on evidence suggesting that '70–75% of internet searches worldwide are done through Google',[34] Justice Fenlon was content that blocking the relevant content on Google would ensure that 'the defendants will not be commercially successful'.[35] In other words, in this context, she was prepared to accept

[30] For a detailed discussion of Olympic broadcasting rights, see T Ryston-Pratt, 'Olympic TV Rights' (2002) 21(4) Comm L Bull 12–15.

[31] M Williams, 'Olympic Games Go for the Gold Online' (on file with author).

[32] At least at the time of speaking, the chairman of the IOC marketing commission and Internet Work Group.

[33] S Klein, 'Back to the 20th Century: The Olympics and the Internet' <http://mason.gmu.edu/~sklein1/index_files/contentexchangecolumns/column11.htm>, accessed 4 April 2017.

[34] *Equustek Solutions Inc v Jack* (2014) BCSC 1063 <http://www.courts.gov.bc.ca/jdb-txt/SC/14/10/2014BCSC1063.htm>, accessed 2 April 2017 [32].

[35] ibid [152].

a 'leakage' of as much as 30 per cent. However, she was not prepared to accept blocking on Google's Canadian site 'google.ca' only, even though it was suggested that around 95 per cent of Canadians access Google search via the Canadian site.

So, let us consider these figures in more detail. If we imagine that Google's market share is at the higher end of the estimate—that is, that it is 75 per cent—then even if only 95 per cent of those users are directed to the Canadian Google site, the leakage is only 28.75 per cent. Thus, within the accepted uncertainty as to Google's exact market share (70–75%), there is also a possibility that a 95 per cent rate of users being blocked out from the relevant content fits within the accepted leakage of up to 30 per cent.

In the end, this mathematical exercise may not matter. After all, why should the law require 100 per cent efficiency in the context of geo-blocking when it does not do so in most other settings? In my view, as noted, where sophisticated geo-location is applied, it is typically accurate enough for legal purposes.

11.6 Concluding Remarks—Retiring from the Ranks of the Doomsday Prophets

This chapter ought to have demonstrated that geo-location technologies are important tools in our pursuit of the goal of jurisdictional interoperability. But they are tools akin to how dynamite is a useful but dangerous tool for infrastructure developments—we must be careful in how we use such tools. Thus, legislators and courts must recognise the complexity and diversity of geo-location technologies.

Up until now, I could reasonably be criticised as being another failed doomsday prophet based on what I have said about how geo-location will impact the Internet. However, as must be clear from what I have just said, today I retire from that line of work. Yet there was one prediction I got right in 2004. While some others, writing on geo-location in the early days, saw no future or relevance of geo-location, indeed calling it a 'red herring', I observed: 'There is an Indonesian saying where novel objects, events or phenomenon are said to be like "hot, hot chicken poo"—hot when first arriving but soon cooling down and being forgotten and of no significance. I think the above has demonstrated that geo-identification and geo-location technologies are much more than "hot, hot chicken poo".'[36] I now go a bit further. Geo-location is here to stay, geo-location will not cause the Internet's

[36] Svantesson (n 1) 137.

'Ragnarök'[37] and in fact geo-location may be the only way to uphold a multi-faceted and multicultural, yet law-abiding, Internet that remains relatively border-disregarding.

Thus the Internet will inevitably be transformed from a relatively borderless dimension into a medium that recognises geographical and legal borders. Such a development seems particularly unavoidable in the light of current 'effect-focused' conflict of laws rules. In other words, from the perspective of Internet regulation, geo-location technologies may to a large extent have eliminated the regulatory difficulties associated with the Internet's 'borderlessness'—geo-location technologies may place borders on the Internet. Indeed, already in 2005, Reidenberg noted that:

> [w]hile online technologies were initially designed for geographically indifferent access, nothing fixed the technology in stone. Commercial pressures and the dynamic nature of the Internet have resulted in geolocation and the recreation of geographic origin and destination. This design feature and its malleability mean that Internet activity is 'purposely availing' throughout the Internet whenever content is posted without geolocation filtering.[38]

To conclude, the impact that geo-identification, or lack thereof, will have on the jurisdictional issues that may arise in a particular case must be determined in the light of the individual facts of that case. However, I remain convinced that the use, or lack of use, of geo-location technologies must be considered in every analysis of jurisdictional issues involving the Internet. Courts and other lawmakers that fail to do so are, in my view, negligent. And the ultimate decision as to whether we want to pursue a path that will significantly increase the use of geo-location technologies depends on against what we measure such a development. If our option is widespread geo-location or an open Internet, I suspect most of us would prefer the open Internet. However, if our choice is between widespread geo-location or an Internet where only content that does not offend any law anywhere in the world is available, then I certainly favour widespread geo-location.

[37] End of the world in Norse mythology.

[38] JR Reidenberg, 'Technology and Internet Jurisdiction' (2005) 153 U Pa L Rev 1951, 1956.

12

A Doctrine of Selective Legal Compliance

The discussion of the matter of scope of (remedial) jurisdiction brought attention to the vulnerabilities of the Internet intermediaries that are so central to the functioning of the Internet of today. Here, I want to elaborate on that topic and discuss something as provocative as the possibility of a doctrine of selective legal compliance to protect the Internet intermediaries. Of course, Google, Microsoft, Facebook, etc. are corporate entities driven by profit. So why would we want to provide them with any special protection? Well, the reality is that, without them, it is not just shareholders that would suffer; we—the average Internet users—need Internet intermediaries to find the content we want, to store our emails, and to cater for our social interaction. This has already been raised in the Introduction.

Debates about the role of, and possible protection for, Internet intermediaries are often characterised by clashing extremist points of view. The free speech advocates (typically from the US) seek to impose an uncompromising and all-embracing free speech regime, where any restrictions imposed by an Internet intermediary on what is uploaded by Internet users is seen as a gross violation. Such extreme views, while ultimately unhelpful, are understandable from a 'ask for everything to get something' perspective. However, this may be an area in which the saying 'be careful what you wish for' is of particular relevance. After all, an absolute, or near absolute, right to free speech online will clearly clash with the reasonable, indeed necessary, goals of addressing, for example, online bullying, defamation, and child abuse. At the same time, the problem is obvious: as soon as we abandon the ideal of absolute free speech, Internet intermediaries are faced with the difficult, and I suspect unwelcome, task of being the judges and enforcers of 'good taste'—Internet intermediaries become the censors and gatekeepers of speech.[1] This

[1] Google eg has stated that 'Google is not, and should not become, the arbiter of what does and does not appear on the web. That's for the courts and those elected to government to decide', Rachel Whetstone, 'Free expression and controversial content on the web' (14 November 2007) <http://googleblog.blogspot.com/2007/11/free-expression-and-controversial.html>, accessed 4 April 2017.

Solving the Internet Jurisdiction Puzzle. First Edition. Dan Jerker B. Svantesson.

is a role for which Internet intermediaries are typically ill-suited. Indeed, we may legitimately question whether society should assign such a crucial role to private entities with the types of agenda normally held by private entities. Some would here hurriedly point to the fact that newspapers, as well as radio and TV broadcasters, have for a long time acted in the role of censors in deciding what content to make available. But the role of the Internet intermediary is so fundamentally different that we cannot, and should not, engage in an uncritical comparison with such media outlets. Indeed, the argument is often made that Internet intermediaries are more like the postal service, passively distributing other people's content without interference. To avoid stepping into the quagmire of analogies, I will here simply observe that no previous intermediaries in the history of the mankind have been faced with the scale of user-generated content with which Internet intermediaries are faced. In my view, this difference in quantity amounts to a difference in character, and the role of Internet intermediaries must be approached with fresh eyes, free from the contamination of preconceived notions based on comparisons with the roles of other intermediaries. Now to the other side of the coin. Just as the views of the free speech extremist may be hard to accept, so are the extremist views in the legal compliance camp. The problem is this: while most people would expect Internet intermediaries to abide by the law of their respective countries, they would probably not wish for Internet intermediaries to abide by all laws of all other countries in the world. In the end, such compliance would lead to Internet intermediaries being forced to take account only of the most restrictive laws from all the countries in the world. Such a race to the bottom is doubtless an unhealthy direction for the Internet. And if this is not what we want, we need to consider whether a globally active Internet intermediary can ever be excused for not complying with all the laws around the world that claim to apply to its conduct. Is it time to construct *an international law doctrine of selective legal compliance*? If we answer that question in the affirmative, what features ought such a doctrine to have and how should globally active Internet intermediaries assess which laws they abide by? These are, at least to a degree, novel and unique questions in international law.

12.1 To Comply or not to Comply? That Can Never Be the Question

As noted in the Introduction, it is of course already the case that Internet intermediaries enjoy certain specific forms of protection such as the aforementioned §230 of the US *Communications Decency Act* of 1996 and Articles

12–15 of the E-Commerce Directive.[2] Nevertheless, two typically central pillars in any discussion of law are (1) that the law applies equally to everyone and (2) that legal compliance is mandatory as opposed to discretionary. In the light of this, it may be difficult to even begin to imagine any form of a doctrine of selective legal compliance—a doctrine legitimising the failure to comply with some otherwise applicable law. Can it be accepted that globally active Internet intermediaries are only expected to comply with some of the laws around the world that purport to govern their conduct? Is such a doctrine in fact a necessity in the pursuit of the jurisdictional interoperability advocated in this book? A wealth of arguments may be presented for and against a strict insistence on globally active Internet intermediaries complying with all laws around the world. I will here only focus on some such arguments.

First of all, we may consider the burden involved in such compliance. A globally active Internet intermediary would need to consider the private international law rules of every state on the planet as well as the substantive laws of all those states. This is a heavy burden indeed. In fact, one would imagine that this burden may represent an unsurmountable obstacle for any Internet start-up, at least for those aiming to be sure they do not violate any laws whatsoever. Thus, insistence on strict compliance with all the laws worldwide may be anticompetitive as only the largest companies stand any chance at all of coping with the burden of such compliance.

Second, as hinted at above, much of the law Internet intermediaries may have to comply with forces them to make judgement calls about whether to allow or to remove content posted by third parties. The valuable work of Sartor has highlighted that placing Internet intermediaries in such a position may lead to 'collateral censorship'.[3] In fact, as Sartor correctly points out, the problem may be such that Internet intermediaries are tasked with deciding whether certain content is legal or illegal where it is questionable whether there are objective standards conclusively determining that matter.[4] Arguably, in any such situation we are setting Internet intermediaries up to fail—they are, as the saying goes, 'damned if they do and damned if they don't'.

Finally, we need to confront the complicated issue of 'true conflicts' as distinct from 'false conflicts'. One often sees the adherence to the harshest

[2] Directive (EC) 2000/31 of the European Parliament and Council, 8 June 2000, on Certain Legal Aspects of Information Society Services, in particular Electronic Commerce [2000] OJ L178/1, 369.

[3] Giovanni Sartor, 'Provider's Liability and the Right to Be Forgotten' in Dan Svantesson and Stanley Greenstein (eds), *Nordic Yearbook of Law and Informatics 2010–2012: Internationalisation of Law in the Digital Information Society* (Ex Tuto 2013) 101–37, 111.

[4] ibid.

rules as a proposed solution to the difficulty of variances in legal standards where more than one standard applies to specific conduct. Such suggestions rely on notions such as that articulated by Justice Souter, that '[n]o conflict exists, [...] where a person subject to regulation by two states can comply with the laws of both.'[5]

As I have stressed elsewhere, I object to this duties-focused approach since it tells only part of the story. Essentially what Justice Souter and others are saying is that we should focus only on the duties imposed by law. If the *duties* do not conflict, the laws do not conflict. This is preposterously simplistic a perspective. It completely neglects the importance of the *rights* that laws provide. Importantly, the correlative relationship between rights and duties to which we may be accustomed from a domestic law setting does not necessarily survive when transplanted into a cross-border environment; that is, rights provided under one state's legal system may not necessarily create corresponding duties under other legal systems.

I argue that in assessing whether two (or more) laws are in conflict we need to take account of both the duties and the rights for which those laws provide. In other words, even where the duties do not clash, the rights of one country may clash with the duties of another state.

The difference can be illustrated by way of an example. Imagine that the laws of State A specifically provide for a right of religious freedom, while the laws of State B specifically impose a duty of adherence to Norse 'pagan' faith. Where a person, for one reason or another, finds herself bound to comply with both the laws of State A and those of State B, there is no conflict in the view of the reasoning put forward by Justice Souter and others—such a person can comply with the law of both states by adhering to Norse pagan faith.

In contrast, from the perspective I advocate here, there is a conflict since the right provided by the law of State A cannot be freely exercised while at the same time complying with the duty imposed by the law of State B (except, of course, by those who voluntarily choose to exercise their right to worship Odin, Tyr, Thor, Freja, etc.). In the light of this, I argue that calls for compliance with the strictest rules, as a solution to the problem of conflicting laws, are misguided. To this we may add, should it be necessary, that ensuring compliance with the strictest law is not an as easy task as the naive proponents may think. After all, to assess which law is the strictest we need to assess all laws. Further, it is of course not the case that one state's laws will be the strictest in all respects. Thus, the strictest law will be made up of a patchwork of norms from various legal systems.

[5] WS Dodge, 'Extraterritoriality and Conflict-of-Laws Theory: An Argument for Judicial Unilateralism' (1998) 39(1) Harv Intl LJ 101, 136.

Taken together, these points represent a reasonably strong case for an international law doctrine of selective legal compliance allowing globally active Internet intermediaries to avoid the burden of having to comply with all the laws of all the states in the world. At the same time, there is no lack of arguments that can be presented against such a doctrine.

One of the most compelling arguments for insisting on globally active Internet intermediaries having to comply with the law of all the states they come into contact with is found in the fact that they could be said to have chosen to operate globally. It may then not be unreasonable to ask them to comply globally. Indeed, to simplify somewhat, given that companies operating solely in State A or solely in State B have to abide by the laws of States A and B respectively, why should a company avoid the laws of State A and State B merely because it operates in both states? As this example illustrates, the absence of an insistence on strict compliance with all the laws worldwide may also be anticompetitive.

Another strong reason for insisting on globally active Internet intermediaries having to comply with the law of all the states they come into contact with is that the idea of certain entities being beyond, or even above, the law could be seen to undermine the law as an institution, and represent a departure from long-standing tradition necessary for a functioning society. Here we need to distinguish between a party being *beyond* the law and a party being *ab*ove the law. Even legal theorists who emphasise 'obedience' accept that law can be law even where its enforcement is plagued by leakage.[6] In fact, it may be said that lacking enforcement is characteristic of cross-border legal issues. Thus, the occurrence of parties being beyond the reach of the law is nothing new and does not necessarily undermine law as a foundational institution in society. The idea, however, that some parties should be above the law is harder, if not impossible, to digest.

12.2 Features of a Doctrine of Selective Legal Compliance

While potentially offensive at first sight, selective legal compliance is, of course, already happening on a practical level in that global and multinational

[6] See eg Kelsen's observation that '[w]ithin a legal order which as a whole is efficacious there may occur isolated norms which are valid and which yet are not efficacious, that is, are not obeyed and not applied even when the conditions which they themselves lay down for their application are fulfilled.' (Hans Kelsen, *General Theory of Law and State* (Anders Wedberg tr, Lawbook Exchange 2011) 119.)

companies in particular have to select which states' laws they seek to abide by. It is naive to think of legal compliance in traditional terms such as which law applies to a particular activity. In an era of hyper-regulation our thinking must be focused on risk assessment and risk minimalisation. Here, I will investigate what should guide a globally active Internet intermediary in deciding which applicable laws it complies with.

A starting point here must be that Internet intermediaries ought not to be put in a position where compliance with one state's laws necessitates the violation of another state's laws. Such conflicts are unfruitful, harmful, and to be avoided. Importantly, examples of provisions aimed at avoiding such clashes can be found. Looking, for example, at Australia's *Privacy Act 1988* (Cth), section 6A limits the extraterritorial effect of that Act by providing that: '[a]n act or practice does not *breach* an Australian Privacy Principle if: (a) the act is done, or the practice is engaged in, outside Australia and the external Territories; and (b) the act or practice is required by an applicable law of a foreign country.'[7] Similar measures could be implemented more broadly. And where they cannot ensure that the type of clashes described here are eliminated, the option of providing Internet intermediaries with other forms of protection—such as so-called 'clawback' statutes—should be considered.

Furthermore, it seems beyond intelligent dispute that a globally active Internet intermediary must abide by the law of those states in which it wishes to have an ongoing, systematic and substantive presence, whether that presence is physical or virtual. It can be expected that business interests will inevitably impact the company's decisions in this context. If a globally active Internet intermediary, for example, wishes to have an ongoing, systematic, and substantive presence on the European market, it must accept the burden of complying also with the more onerous aspects of EU law. However, as was discussed in more detail earlier, where the Internet intermediary's contact with a particular jurisdiction is not ongoing, systematic, and substantive, states must be more realistic about what aspects of their laws a foreign party is expected to abide by, and the type of 'layered approach' I advocate in Chapter 10 ought to be applied.

Additionally, globally active Internet intermediaries ought to develop clear and transparent internal guidelines for how they decide whether to abide by foreign law in states in which they do not have an ongoing, systematic, and substantive presence. Articulating the details of such internal guidelines quickly takes one into moral philosophical territory beyond the scope of this discussion. However, corporate values and articulated attitudes towards Corporate Social Responsibility ought to be useful tools. For example, globally active

[7] *Privacy Act 1988* (Cth) s 6A(4).

Internet intermediaries should—to the extent they do not already do so—incorporate matters such as the protection of children and the prevention of online bullying etc. into their Corporate Social Responsibility agendas. Such goals ought to be uncontroversial in large parts of the world. They may also include more controversial matters such as the general promotion of democratic values. Furthermore, such guidelines may appropriately place reliance on principles found in international human rights law such as the International Covenant on Civil and Political Rights[8] and the European Convention for the Protection of Human Rights and Fundamental Freedoms.[9]

In addition, one could picture Internet intermediaries appointing sophisticated panels made up of highly respected former judges from the highest courts, as well as particularly qualified academics, lawyers, and policy-makers who are tasked with guiding the company's application of the mentioned internal guidelines. Such panels could be set up individually for each Internet intermediary, or potentially as an independent body, as a pooled resource for a group of Internet intermediaries. The decisions of such panels could, where appropriate with regard to matters such as privacy, be made public so as to provide transparency and guidance for future matters.

In an interesting article, Pauwelyn analyses why the World Trade Organization (WTO) places increasing reliance on expert advice in its dispute settlement. Five reasons are presented:

> First, WTO agreements themselves became more technical, both in the trade/economic sense […] and the factual/scientific sense […]. Secondly, a number of WTO obligations adopt an explicit economic/scientific criterion of legality. […]. Thirdly, WTO dispute settlement has been 'legalised': panels have compulsory jurisdiction, panel reports are virtually automatically adopted and the legal aspects of a dispute are subject to review by an Appellate Body. […]. The new rules-based process has increased the number of reluctant defendants as well as the incentive to dispute the facts. Hence, the need to bring in neutral experts. Fourthly, and linked mainly to the phenomenon of *amicus curiae* briefs, the stakes at play in WTO disputes are no longer limited to government-to-government trade concessions. WTO rules have a direct impact also on individual economic operators, including consumers and citizens at large. This means that WTO member-governments need to explain the positions they take before a WTO panel, as well as the outcome finally obtained, to an ever wider domestic audience. This may provide another incentive for 'aggressive litigation'. Another consequence of this 'indirect effect' of WTO law is that affected private parties—which are perhaps not always fully heard by their

[8] International Covenant on Civil and Political Rights, opened for signature 16 December 1966, 99 UNTS 171 (entered into force 23 March 1976).

[9] Convention for the Protection of Human Rights and Fundamental Freedoms, as amended by Protocol Nos 11 and 14 (Rome, 4 XI 1950).

governments—increasingly want to express their (expert) opinion directly before the panel. Fifthly, expert advice is keenly sought after and accepted by panels because of the important role it can play in the legitimising of WTO decision-making.[10]

I suspect all of these reasons could equally be put forward to support the type of panel(s) I am advocating here. Further on this topic, such panels could deal with both proactive questions, such as whether or not abide by a particular foreign law, and reactive questions, such as how to respond to applications to have certain content blocked/removed. Finally on this topic, one can imagine a system under which an Internet intermediary that has submitted a controversial matter to such a panel, and acted according to the panel's recommendation, would be protected from liability. Indeed, such a scheme could even be linked to an insurance arrangement. However, pursuing the details of this matter here would take us too far afield.

In addition to the aforementioned, we may be able to use the following set of factors in assessing whether an Internet intermediary may be justifiably protected by a doctrine of selective legal compliance, in a specific situation:

1. To what extent was the law being applied communicated to the Internet intermediary against whom the law is being applied?
2. To what extent was the law being applied understandable to the Internet intermediary against whom the law is being applied?
3. To what extent is the law being applied consistent with generally shared views of right and wrong?
4. To what extent is the law being applied consistent with the laws of the Internet intermediary's country/countries of operation?
5. To what extent was the law being applied certain and predictable?
6. To what extent is there a connection between the country where the action is being taken and the conduct of the Internet intermediary?[11]
7. To what extent was the law being applied likely to clash with competing legitimate interests?

It seems to me that where a scenario is scoring low on all, or some, of the first six of these parameters, we may be witnessing an illegitimate expansion of domestic laws onto Cyberspace, and as a result, it may be appropriate to extend to the Internet intermediary the protection afforded under a doctrine of selective legal compliance. The same would be true where compliance with

[10] Joost Pauwelyn, 'The Use of Experts in WTO Dispute Settlement' (2002) 51 Intl & Comp LQ 325, 326 (footnotes omitted).

[11] While modified, these factors are largely based on research presented in: Dan Svantesson, 'The Holy Trinity of Legal Fictions Undermining the Application of Law to the Global Internet' (2015) 23(3) Intl JL & Info Tech 219–34.

the relevant law would negatively impact competing interests to an unacceptable degree. In addition, it seems reasonable to take account of the scope of jurisdiction claimed by the original authority as part of deciding whether intermediary should collaborate or not.

In the case that this is viewed as providing too broad a protection for Internet intermediaries, a cut-down version can also be imagined under which an Internet intermediary is afforded the protection of the doctrine of selective legal compliance only where: (1) the content is not produced by the intermediary, (2) the content is lawful under the laws of the country where protection is sought, (3) the content is only unlawful in a small number of states, (4) the intermediary is asked to remove/block/delist the content globally. The point is that it is possible to imagine any number of versions of a doctrine of selective legal compliance once the idea as such has been accepted as necessary, or at least useful.

At any rate, in writing what has gone before, I have not failed to recognise how it undermines the fundamentals of the legal system—a system that, despite its warts and blemishes, corruption and distortions, is aimed at creating order where there otherwise may be none. Thus, despite the logical conclusion reached if we follow the reasoning so far in its full course, I am not advocating that we revert to a pre-law caveman landscape. I am not seeking to create total anarchy. I am not even a worshipper at the altar of Cyber libertarianism. However, as courts and legislators increasingly make claims to regulate online conduct, they must never forget the fairy-tale-like fiction the legitimacy of their rules rely upon when extended too far into Cyberspace.

12.3 Concluding Remarks Regarding Selective Legal Compliance

This chapter discusses the important role Internet intermediaries play for the proper functioning of the Internet of today. It elaborates on this topic and discusses the possibility of a doctrine of selective legal compliance to protect the Internet intermediaries. The chapter demonstrates that the law already takes steps towards protecting Internet intermediaries due to their unique role and argues that expanding this protection into the mentioned doctrine may be a surprisingly natural development—and a development that would be most helpful for establishing jurisdictional interoperability.

One aspect that is not pursued is how such a doctrine may be introduced. Several possibilities are available ranging from multilateral agreements to domestic law initiatives. That is, however, a matter that I will not pursue here.

13

Final Remarks

According to legend, Voltaire—on his deathbed—was asked to renounce Satan. As the story goes, he calmly responded 'Now, now my good man, this is no time to be making enemies.'[1] A somewhat similar reasoning may be applied to the concluding chapter of a book. If a reader has come this far, the author ought not advance anything challenging, anything that may offend the brave reader's sensibilities. I have already—in the preceding chapters—renounced territoriality and advanced a territoriality nihilistic alternative jurisprudential framework for jurisdiction. In the context of jurisdiction, territoriality essentially is meant to fulfill two functions, and it fails at both. The first is that territoriality is meant to be a criterion for when a state can claim jurisdiction. But especially online it is too easy to find territorial anchor-points for jurisdictional claims. Thus, as a hurdle for excessive jurisdictional claims, territoriality does not work well as it is too easy to satisfy. The second function of territoriality is that it is meant to act as a stop sign providing a warning when you are entering the exclusive domain of another state. But again, territoriality fails since it is simply unrealistic to think that a state will be connected to the global community and still enjoy traditional notions of territorial exclusiveness.

I have rejected the convention of categorising jurisdiction as being prescriptive, adjudicative, or enforcement jurisdiction, and I have argued that instead the reasonableness and legitimacy of any given jurisdictional claim may be assessed by reference to the degree to which it interferes with the sovereignty of another state. In the pursuit of what I termed jurisdictional interoperability, I have also called for the introduction of several new concepts, such as the contextual legal system, investigative jurisdiction, scope of (remedial) jurisdiction, the layered approach, and market sovereignty based on market destroying measures. Further, I have advanced the idea of a doctrine of

[1] <http://www.goodreads.com/quotes/30291-now-now-my-good-man-this-is-no-time-to>, accessed 4 April 2017.

Solving the Internet Jurisdiction Puzzle. First Edition. Dan Jerker B. Svantesson.

selective legal compliance to protect the Internet intermediaries and pointed to the role geo-location technologies can play. In addition, I have discussed what law is, and argued that we can say that, quite apart from being a tool to decide legal disputes and to provide a framework to control, guide, and plan life out of court, law is a tool to communicate the values of the society that created the law. In addition, I have described the difference between bite jurisdiction and bark jurisdiction and I have argued that our interpretation of law should be guided by: 'fairness' (defined here as the search for the most appropriate resolution of the dispute at hand), 'consequence focus' (defined here as the search for the option having the most favourable consequences for the future), and 'harmonisationalism' (defined here as the search for the option that is most clearly in line with international best practice).

Were all these proposals to be accepted, it would mean a total reform of how we deal with Internet jurisdiction, and I realise that it is not a small thing to ask the reader to accept all this. Thus, here I will ask for nothing more; I will merely summarise a few key points and seek again to draw attention to the significance of the task that lies ahead: the task of solving the Internet jurisdiction puzzle.

First, however, I wish to offer an attempt at outlining a guiding principle – an overarching doctrine – to guide our work on solving the Internet jurisdiction puzzle. And doing so will take me back to the tables of the longhouses in Viking-era Scandinavia.

There is a word said to be quite unique to the Swedish language. The word *lagom* means 'just enough' or 'just right'. At least according to folklore, it stems from the phrase *laget om* ('around the team') from the Viking tradition of drinking enough when the drinking horn was passed around, without drinking so much that there is not enough for everyone.

I think the concept of *lagom* is apt indeed to describe how we must approach the Internet jurisdiction puzzle. Most obviously, neither excess nor abstinence are acceptable paths forward; that is, emptying the drinking horn before everyone has had a chance to get their fair share would be an insult, but a refusal to take part in the drinking would be equally insulting. Similarly, states should not make excessive jurisdictional claims, but equally well they should not decline to exercise jurisdiction where doing so is called for. Further, the *lagom* doctrine incorporates a context-specific proportionality. If the drinking horn is large, or the group of people sharing it small, each member can drink more than if the proportions are in the reverse. In the same manner, jurisdictional claims (and their scope) need to be adjusted to the context. However, the comparison goes further than that. In fact, it is possible to link numerous international law concepts to the *lagom* doctrine. Consider the concept of 'comity' that clearly can be seen in the requirement of not drinking so much that there is not enough for everyone. Or why not the 'due diligence' requirement that states must ensure that other states' rights and interests are not violated due to activities over which the first state has jurisdiction; whether we are talking about Viking-era drinking or about jurisdiction, everyone must partake and claim their share.

In the light of the above, perhaps it can be said to be the case that our international law principles on jurisdiction are no more advanced than was the Viking-era drinking etiquette? And perhaps they do not need to be—perhaps the *lagom* doctrine is the answer—or at least a component of the answer—to the Internet jurisdiction puzzle?

At any rate, in November 2016, at the first Global Internet and Jurisdiction Conference of the multistakeholder policy network Internet & Jurisdiction, former Swedish Prime Minister Carl Bildt expressed the view that 'how the jurisdiction challenges on the Internet are addressed will define our digital future'.[2] This is no exaggeration. The risk are real and ignoring them will come at a severe cost. Put simply, we need to act now, and indeed we are already somewhat late to the party.

A Report summarising that same conference notes: 'The participants recognized that no actor or stakeholder group can solve these challenges alone and that uncoordinated efforts or inaction come with a high cost for the future of the digital economy, human rights, and cybersecurity.'[3] To achieve coordination surely the first step must be to generate a common understanding of jurisdiction. Because as this book must have made clear, there is no such common understanding at present (even though there is no lack of commentators that—for reasons ranging from ignorance to cynicism—argue that there is such a common understanding). Regrettably, discussions about Internet jurisdiction are constantly muddled due to, sometimes intentional and sometimes unintentional, overstatements of legal positions sparked by differences in culture, societal values, legal points of view, and, most importantly, different interests. I predict that we are currently heading towards an even greater degree of difference in culture, societal values, legal points of view, and different interests. We will see more and more value clashes as more and more countries of diverse background are becoming more and more active online.

[2] Summary by the Internet & Jurisdiction Secretariat: Global Internet and Jurisdiction Conference <http://www.internetjurisdiction.net/uploads/pdfs/GIJC-Secretariat-Summary.pdf> 2, accessed 4 April 2017.

[3] ibid.

By questioning the assumption of a *current* common understanding of the relevant concepts, and indeed by questioning the suitability of many commonly adhered to concepts, I have tried to take us in the direction of a *future* common understanding of jurisdiction. I have also sought to lay the groundwork for a workable path forward by outlining a jurisprudential framework for jurisdiction—a framework that moves away from dysfunctional territoriality—moreover, a framework that merges public and private international law as far as jurisdiction is concerned.

I have emphasised that, while novel in some respects, the framework I advance rests on solid and well-established principles deeply engraved in both public and private international law; thus, the principles identified through my 'kitchen sink experiment' can be seen to have quite a regal lineage indeed.

In many ways, my proposed paradigm change is unthreatening because it does not preclude the continued use of territoriality where territoriality works. The only change in that setting is that territoriality is no longer a core principle but rather a second level principle that arguably is more easily departed from via exceptions.

Despite the many practical measures I have advanced, I acknowledge that the discussion in this book is rather theoretical in nature. As such, it may be criticised as providing theoretical solutions to practical problems. Such criticism, however, is ill-conceived, hints at ignorance, and lacks actual sting. Upon a sober-minded consideration of the real state of things, it is obvious that practical solutions must also rest on theoretically sound foundations. After all, as soon as a practical solution is faced by what has been termed 'hard cases', the proper application of that practical solution depends on its theoretical underpinnings. And it is my hope that this book goes some way towards constructing a sound and solid foundation for the practical solutions we need to address the Internet jurisdiction puzzle.

What we are trying to do here is to build practical structures to address real world practical issues. My aim is mainly to ensure that these structures are built on a solid foundation, one that will allow them to work well and to withstand the test of time. Territoriality is—quite clearly—already past its use-by date. And especially in areas that move forward at a fast pace, we cannot constantly and only keep our eyes on the rear-view mirror of cases such as *Lotus*; we need to look forward to where we are going, or we will drive off the road and crash.

To conclude, as noted in the Introduction, this is a book *about* solving the Internet jurisdiction puzzle. Like a cookbook, it has presented a recipe, but of course not the dish itself. I hope that, like a really good cookbook, the recipe presented has increased the reader's appetitive for the dish in question—in this case, the solution to a serious legal problem with far-reaching, global, implications. At the same time, I admit that the recipe has been kept rather vague in some parts, perhaps some ingredients are still missing and the more detailed cooking instructions are rather political in nature and therefore outside the scope of this work.

As I put the finishing touches to this monograph, I can already hear the type of calls not to 'throw out the baby with the bathwater' that (too) often accompany any call for substantial reform. In reply to such calls, I would argue that if the territoriality principle is the 'bathwater', it has now become cold, contaminated, and unpleasant, and there is little option but to throw it out. And if the proverbial baby represents some form of world order created by the focus on territoriality, I argue that the baby has left the bath a long time ago. Indeed, I would go as far as to say that the baby has now grown up and is better seen as a rebellious teenager with little or no interest in bathing.

Bibliography

BOOKS

Bentham, Jeremy, *Of Promulgation of the Laws and Promulgation of the Reasons Thereof* (1843) < http://www.laits.utexas.edu/poltheory/bentham/promul/index.html>, accessed June 2017.

Berman, Paul Schiff, *Global Legal Pluralism: A Jurisprudence of Law beyond Borders* (CUP 2014).

Bygrave, Lee A, *Data Privacy Law: An International Perspective* (OUP 2014).

Bygrave, Lee A, *Internet Governance by Contract* (OUP 2015).

Collins, Matthew, *Collins On Defamation* (OUP 2014).

Colonna, Liane, *Legal Implications of Data Mining: Assessing the European Union's Data Protection Principles in Light of the United States Government's National Intelligence Data Mining Practices* (Ragulka 2016).

Crawford, James, *The Creation of States in International Law* (2nd edn, OUP 2006).

Crawford, James, *Brownlie's Principles of Public International Law* (8th edn, OUP 2012).

Davies, M, A Bell, and P Brereton, *Nygh's Conflict of Laws in Australia* (9th edn, LexisNexis Australia 2014).

European Max Planck Group on Conflict of Laws in Intellectual Property, *Conflict of Laws in Intellectual Property: The CLIP Principles and Commentary* (OUP 2013).

Freeland, Steven, *Addressing the Intentional Destruction of the Environment During Warfare Under the Rome Statute of the International Criminal Court* (Intersentia 2015).

Fuller, Lon L, *The Morality of Law* (2nd edn, Yale University Press 1969).

Gillies, LE, *Electronic Commerce and International Private Law: A Study of Electronic Consumer Contracts* (Ashgate 2008).

Goldsmith, Jack L, and Eric A Posner, *The Limits of International Law* (OUP 2005).

Goldsmith, Jack, and Tim Wu, *Who Controls the Internet? Illusions of a Borderless World* (OUP 2008).

Grahn, Jan-Olof, *Om Svensk Underrättelsetjänst* (Medströms Bokförlag 2016).

Guzman, Andrew T, *How International Law Works: A Rational Choice Theory* (OUP 2008).

Handl, Günther, Joachim Zekoll, and Peer Zumbansen (eds), *Beyond Territoriality* (Martinus Nijhoff 2012).

Hart, HLA, *The Concept of Law* (3rd edn, OUP 2012).

Ireland-Piper, Danielle, *Accountability in Extraterritoriality* (Edward Elgar 2017).

Janis, Mark Weston, *International Law* (7th edn, Wolters Kluwer 2016).

Kelsen, Hans, *General Theory of Law and State* (Transaction 2005).

Kelsen, Hans, *General Theory of Law and State* (Anders Wedberg tr, Lawbook Exchange 2011).

Keyes, Mary, *Jurisdiction in International Litigation* (The Federation Press 2005).

Kohl, U, *Jurisdiction and the Internet: Regulatory Competence over Online Activity* (CUP 2007).

Koskenniemi, Martti, *The Politics of International Law* (Hart 2011).

Lauterpacht, Hersch, *The Function of Law in the International Community* (OUP 2011).

Lessig, L, *Code and Other Laws of Cyberspace* (Basic Books 1999).

Lewis, Sir George Cornewall, *On Foreign Jurisdiction and the Extradition of Criminals* (John W Parker & Son 1859).

Lobato, Ramon, and James Meese (eds), *Geoblocking and Global Video Culture* (Institute of Network Cultures 2016).

MacCormick, Neil, *Institutions of Law* (OUP 2007).

Neff, SC (ed), *Hugo Grotius on the Law of War and Peace* (CUP 2012).

Nowak, M, *UN Covenant on Civil and Political Rights* (NP Engel, 1993).

Oster, Jan, *European and International Media Law* (CUP 2017).

Palfrey, John, and Urs Gasser, *Interop: The Promise and Perils of Highly Interconnected Systems* (Basic Books 2012).

Price, James, and Felicity McMahon (eds), *Blackstone's Guide to The Defamation Act 2013* (OUP 2013).

Putnam, Tonya L, *Courts Without Borders: Law, Politics, and U.S. Extraterritoriality* (CUP 2016).

Reed, Chris, *Making Laws for Cyberspace* (OUP 2012).

Ross, A, *On Law and Justice* (Stevens & Sons 1958).

Ryngaert, Cedric, *Jurisdiction in International Law* (2nd edn, OUP 2015).

Schmitt, Michael N (ed), *Tallinn Manual on the International Law Applicable to Cyber Warfare* (CUP 2013).

Schmitt, Michael N (gen ed), *Tallinn Manual 2.0 on the International Law Applicable to Cyber Operations* (CUP 2017).

Schultz, Thomas, and David Holloway, *The Principle of Comity in Private and Public International Law* (CUP 2017).

Spang-Hanssen, H, *Cyberspace & International Law on Jurisdiction* (DJØF 2004).

Svantesson, Dan, *Private International Law and the Internet* (Kluwer Law International 2007).

Svantesson, Dan, *Extraterritoriality in Data Privacy Law* (Ex Tuto 2013).

Svantesson, Dan, *Private International Law and the Internet* (3rd edn, Kluwer Law International 2016).

Svantesson, Dan, and Radim Polák, *Information Sovereignty* (Edward Elgar forthcoming 2017).

Tang, ZS, *Electronic Consumer Contracts in the Conflict of Laws* (Hart 2009).

Thirlway, Hugh, *The Sources of International Law* (OUP 2014).

Ubertazzi, Benedetta, *Exclusive Jurisdiction in Intellectual Property* (Mohr Siebeck 2012).

von Linné, Carl, *Systema naturæ per regna tria naturæ, secundum classes, ordines, genera, species, cum characteribus, differentiis, synonymis, locis* (10th edn, 1758).

CHAPTERS IN BOOKS

Baldwin, James, 'King Canute on the Seashore' in William J Bennett (ed), *The Book of Virtues for Young People: A Treasury of Great Moral Stories* (Silver Burdett 1995), 18–19.

Bethlehem, Daniel, 'Comments on Hurrell: The End of Geography?' in James Crawford and Sarah Nouwen (eds), *Selected Proceedings of the European Society of International Law*, vol 3 (Hart 2012).

Bing, Jon, 'Building Cyberspace: A Brief History of Internet' in Lee A Bygrave and Jon Bing (eds), *Internet Governance: Infrastructure and Institutions* (OUP 2009).

Bygrave, Lee, 'Data Privacy Law and the Internet: Policy Challenges' in Normann Witzleb, David Lindsay, Moira Paterson, and Sharon Rodrick (eds), *Emerging Challenges in Privacy Law: Comparative Perspectives* (CUP 2014).

Dorsett, Shaunnagh, 'Mapping Territories' in Shaun McVeigh (ed), *Jurisprudence of Jurisdiction* (Routledge-Cavendish 2007).

Glasze, Georg, 'Geoinformation, Cartographic (Re)Presentation and the Nation State: A Co-Constitutive Relation and Its Transformation in the Digital Age' in Uta Kohl (ed), *The Net and the Nation State—Multidisciplinary Perspectives on Internet Governance* (CUP 2017).

Guastini, R, 'Rule-Scepticism Restated' in Leslie Green and Brian Leiter (eds), *Oxford Studies in Philosophy of Law*, vol 1 (OUP 2011).

Hondius, Ewoud, 'The Unbearable Heaviness of Legal Texts' in Børge Dahl, Thomas Riis, and Jan Trzaskowski (eds), *Liber Amicorum Peter Møgelvang-Hansen* (Ex Tuto 2016).

Hurrell, Andrew, 'International Law 1989–2010: A Performance Appraisal' in James Crawford and Sarah Nouwen (eds), *Selected Proceedings of the European Society of International Law* (Hart 2012).

Khan, Daniel-Erasmus, 'Territory and Boundaries' in Bardo Fassbender and Anne Peters (eds), *The Oxford Handbook of The History of International Law* (OUP 2012).

Knop, Karen, and Susan Marks, 'The War against Cliché: Dispatches from the International Legal Front' in Christine Chinkin and Freya Baetens (eds), *Sovereignty, Statehood and State Responsibility: Essays in Honour of James Crawford* (CUP 2015).

Konow, Berte-Elen R, 'Observations on Some Points of Contact between Private and Public International Law in Sweden' in Patrik Lindskoug et al (eds), *Essays in Honour of Michael Bogdan* (Juristförlaget 2013).

Mann, Francis, 'The Doctrine of Jurisdiction in International Law' in Karl M Meesen (ed), *Extraterritorial Jurisdiction in Theory and Practice* (Kluwer Law International 1996).

McDougal, M, and R Jasper, 'The Foreign Sovereign Immunities Act of 1976: Some Suggested Amendments' in M Landwehr (ed), *Private Investors Abroad—Problems and Solutions in International Business in 1981* (1982).

Michaels, R, 'Territorial Jurisdiction after Territoriality' in PJ Slot and M Bulterman (eds), *Globalisation and Jurisdiction* (Kluwer International 2004).

Nagy, Boldizsár, 'From the National Border to the National Eleven: A (Partial and Partisan) Appraisal of the State System's Performance since the End of the Cold War' in James Crawford and Sarah Nouwen (eds), *Selected Proceedings of the European Society of International Law*, vol 3 (Hart 2012).

Orakhelashvili, Alexander, 'State Jurisdiction in International Law: Complexities of a Basic Concept' in Alexander Orakhelashvili (ed), *Research Handbook on Jurisdiction and Immunities in International Law* (Edward Elgar 2015).

Oster, Jan, 'Which Limits on Freedom of Expression are Legitimate? Divergence of Free Speech Values in Europe and the United States' in Uta Kohl (ed), *The Net and the Nation State—Multidisciplinary Perspectives on Internet Governance* (CUP 2017).

Reed, Chris, 'Cloud Governance: The Way Forward' in Emerging Challenges in Christopher Millard (ed), *Privacy Law: Comparative Perspectives* (OUP 2013).

Ryngaert, Cedric, and Mark Zoetekouw, 'The End of Territory? The Re-Emergence of Community as a Principle of Jurisdictional Order in the Internet Era' in Uta Kohl (ed), *The Net and the Nation State—Multidisciplinary Perspectives on Internet Governance* (CUP 2017).

Sartor, Giovanni, 'Provider's Liability and the Right to Be Forgotten' in Dan Svantesson and Stanley Greenstein (eds), *Nordic Yearbook of Law and Informatics 2010–2012: Internationalisation of Law in the Digital Information Society* (Ex Tuto 2013).

Schmitt, Michael N, and Vihul Liis, 'The Nature of International Law Cyber Norms' in Anna-Maria Osula and Henry Rõigas (eds), *International Cyber Norms: Legal, Policy & Industry Perspectives* (NATO CCD COE 2016).

Simma, Bruno, and Andreas Th. Müller, 'Exercise and Limits of Jurisdiction' in James Crawford and Martti Koskenniemi (eds), *The Cambridge Companion to International Law* (CUP 2012).

Sjöberg, Cecilia Magnusson, 'Rättens internationalisering i det digitala informationssamhället—Folkrätt och rättsinformatik' in Rebecca Stern and Inger Österdahl (eds), *Folkrätten i svensk rätt* (2012).

Smith, Graham, 'Cyberborders and the Right to Travel in Cyberspace' in Uta Kohl (ed), *The Net and the Nation State—Multidisciplinary Perspectives on Internet Governance* (CUP 2017).

Svantesson, Dan, 'Cyberborders through "Code": An All-or-Nothing Affair?' in Uta Kohl (ed), *The Net and the Nation State—Multidisciplinary Perspectives on Internet Governance* (CUP 2017).

Svantesson, Dan, 'Enforcing Privacy Across Different Jurisdictions', in David Wright and Paul De Hert (eds), *Enforcing Privacy: Regulatory, Legal and Technological Approaches* (Springer 2016).

Svantesson, Dan, 'Legal Theories of Private International Law: Overview and Practical Implications for Internet Regulation' in Patrik Lindskoug et al (eds), *Essays in Honour of Michael Bogdan* (Juristförlaget i Lund 2013).

Svantesson, Dan, 'Online Cross-Border Defamation Disputes' in Dan Svantesson and Stanley Greenstein (eds), *Nordic Yearbook of Law and Informatics 2010–2012: Internationalisation of Law in the Digital Information Society* (Ex Tuto 2013).

Svantesson, Dan, 'B2C Sales and Consumer Protection' in Radim Polčák (ed) *Introduction to ICT law—Selected Issues* (Mazarykova univerzita, Brno 2007).

Warf, Barney, 'Alternative Geographies of Cyberspace' in Uta Kohl (ed), *The Net and the Nation State—Multidisciplinary Perspectives on Internet Governance* (CUP 2017).

JOURNALS

Akehurst, Michael, 'Jurisdiction in International Law' (1972–73) 46 BYIL 145.

Bendiek, Annegret, 'Due Diligence in Cyberspace: Guidelines for International and European Cyber Policy and Cybersecurity Policy' (Stiftung Wissenschaft und Politik -SWP- Deutsches Institut für Internationale Politik und Sicherheit, ed, 2016) SWP Research Paper 7/2016 <http://nbn-resolving.de/urn:nbn:de:0168-ssoar-47152-8>, accessed June 2017.

Bertram-Nothnagel, H, 'Enforcement of Foreign Judgments and Arbitral Awards in West Germany' (1976–77) 17 VA J Intl L 393.

Bialostozky, Noah, 'Extraterritoriality and National Security: Protective Jurisdiction as a Circumstance Precluding Wrongfulness' (2014) 52 Colum J Transnatl L 617.

Briggs, Adrian, 'The Principle of Comity in Private International Law' (2012) 354 Recueil des Cours 65.

Brilmayer, L, 'Liberalism, Community and State Borders' (1991) 41 Duke LJ 1.

Brkan, Maja, 'The Unstoppable Expansion of EU Fundamental Right to Data Protection. Little Shop of Horrors?' (2016) 23(5) Maastricht Journal of European and Comparative Law 812.

Brockman-Hawe, Benjamin, 'Using Internet "Borders" to Coerce or Punish: The DPRK as an Example of the Potential Utility of Internet Sanctions' (2007) 25 BU Intl LJ 163.

Burin, FS, 'The Theory of the Rule of Law and the Structure of the Constitutional State' (1965–66) 15 Am U LR 313.

Buxbaum, Hannah L, 'Territory, Territoriality, and the Resolution of Jurisdictional Conflict' (2009) 57 Am J Comp L 631.

Bygrave, Lee, 'Determining Applicable Law Pursuant to European Data Protection Legislation' (2000) 16 Compu L & Sec Rep 252.

Cooper, Daniel, and Christopher Kuner, 'Data Protection Law and International Dispute Resolution' (2017) 382 Recueil des cours 9–174.

Coughlan, S, et al, 'Global Reach, Local Grasp: Constructing Extraterritorial Jurisdiction in the Age of Globalization' (2007) 6 Can JL & Tech 32.

Dodge, WS, 'Extraterritoriality and Conflict-of-Laws Theory: An Argument for Judicial Unilateralism' (1998) 39(1) Harv Intl LJ 101.

Dworkin, Ronald, 'A New Philosophy for International Law' (2013) 41 Phil & Pub Aff 2.

Ford, Richard T, 'Law's Territory (A History of Jurisdiction)' (1998–99) 97 Mich L Rev 843.

Geist, Michael, 'Is There a There There? Towards Greater Certainty for Internet Jurisdiction' (2001) 16 Berkeley Tech LJ 1345.

Gerber, David J, 'Beyond Balancing: International Law Restraints on the Reach of National laws' (1984) 10 Yale J Intl L 185.

Ginsburg, Tom, 'Comment on Svantesson, A New Jurisprudential Framework for Jurisdiction' (2015) 109 AJIL Unbound 86.

Goldsmith, Jack, 'Against Cyberanarchy' (1998) 65 U Chicago L Rev 1250.

Goldsmith, Jack, 'Unilateral Regulation of the Internet: A Modest Defence' (2000) 11 Eur J Intl L 135.

Goldsmith, Jack, and AO Sykes, 'The Internet and the Dormant Commerce Clause' (2001) 110 Yale LJ 812.

Heine-Ellison, Sofie, 'The Impact and Effectiveness of Multilateral Economic Sanctions; A Comparative Study' (2001) 5 Intl J Hum Rights 81, 83.

Hixson, Kathleen, 'Extraterritorial Jurisdiction under the Third Restatement of Foreign Relations Law of the United States' (1988) 12 Fordham Intl L J 127.

'Introductory Comment to the Harvard Draft Convention on Jurisdiction with Respect to Crime 1935' (1935) 29 Supp AJIL 443.

Ireland-Piper, Danielle, 'Extraterritorial Criminal Jurisdiction: Does the Long Arm of the Law Undermine the Rule of Law?' (2012) 13 Melb J Intl L 1.

Ireland-Piper, Danielle, 'Prosecutions of Extraterritorial Criminal Conduct and the Abuse of Rights Doctrine' (2013) 9 Utrecht L Rev (2013) <http://www.utrecht-lawreview.org/index.php/ulr/article/download/243/237>, accessed June 2017.

Jaroszek, Agata, 'European Online Marketplace—New Measures for Consumer Protection against "Old Conflict of Laws Rules"' (2015) 9 Masaryk UJL & Tech 23.

Jaeger-Fine, Desiree, et al, 'Internet Jurisdiction: A Survey of German Scholarship and Cases' [30 June 2013] Fordham Law Legal Studies Research Paper No 2309575 <http://ssrn.com/abstract=2309575> or <http://dx.doi.org/10.2139/ssrn.2309575>, both accessed June 2017.

Jensen, JM, 'Personal Jurisdiction in Federal Courts over International E-Commerce Cases' (2006–07) 40 Loy LAL Rev 1507.

Johnson, David, and David Post, 'Law and Borders—The Rise of Law in Cyberspace' (1996) 48 Stan L Rev 1367.

Kassan, Shalom, 'Extraterritorial Jurisdiction in the Ancient World' (1935) 29 AJIL 237.

Kent, Gail, 'Sharing Investigation Specific Data with Law Enforcement—An International Approach' (Stanford Public Law Working Paper, 14 February

2014) <http://ssrn.com/abstract=2472413> or <http://dx.doi.org/10.2139/ssrn.2472413>, both accessed June 2017.

King, KF, 'Personal Jurisdiction, Internet Commerce, and Privacy: The Pervasive Legal Consequences of Modern Geolocation Technologies' (2011) 21 Albany L J Sci & Tech 61.

Kuner, Christopher, 'Data Protection Law and International Jurisdiction on the Internet (Part 1)' (2010) 18(2) Intl JL & Info Tech 176.

Kuner, Christopher, 'Data Protection Law and International Jurisdiction on the Internet (Part 2)' (2010) 18(3) Intl JL & Info Tech 227.

Kuner, Christopher, 'Extraterritoriality and International Data Transfers in EU Data Protection Law' (2015) 5(4) Intl Data Privacy L 235.

Kuner, Christopher, The Internet and the Global Reach of EU Law (1 February 2017) forthcoming in the *Collected Courses of the Academy of European Law* (OUP) <https://ssrn.com/abstract=2890930>, accessed June 2017.

Lessig, L, 'The Zones of Cyberspace' (1996) 48 Stan L Rev 1405.

Lessig, L, 'The Law of the Horse: What Cyberlaw Might Teach' (1999) 113 Harv L Rev 501.

Levin, Dave, et al, 'Alibi Routing' (2015) Sigcomm <https://alibi.cs.umd.edu/papers/alibi_sigcomm15.pdf>, accessed June 2017.

Lowell, A Lawrence, 'The Limits of Sovereignty' (1888–89) 2 Harv L Rev 70.

Maier, Bernhard, 'How Has the Law Attempted to Tackle the Borderless Nature of the Internet?' (2010) 18 Intl JL & Info Tech 161.

Maier, Harold G, 'Remarks' (1990) 84 Proc Am Soc Intl L 339.

McConnaughay, PJ, 'Reviving the "Public Law Taboo" in International Conflict of Laws' (1999) 35 Stan J Intl L.

Man, CS, 'Conflict in Canengus: The Battle of Consequentialism and Deontology' (2010) 7 J Juris 535.

Meessen, Karl M, 'Conflicts of Jurisdiction under the New Restatement' (Summer 1987) 50 L & Contemp Probs 47.

Menthe, DC, 'Jurisdiction in Cyberspace: A Theory of International Spaces' (1998) 4 Mich Telecom & Tech L Rev 69.

Mills, Alex, 'Rethinking Jurisdiction in International Law' (2014) 84 Brit YB Intl L.

Moerel, Lokke, 'The Long Arm of EU Data Protection Law: Does the Data Protection Directive Apply to Processing of Personal Data of EU Citizens by Websites Worldwide?' (2011) 1 Intl Data Priv L 23.

Morris, Madeline, 'High Crimes and Misconceptions: The ICC and Non-Party States' (2001) 64 L & Contemp Probs 13.

Muir Watt, Horatia, 'A Private (International) Law Perspective: Comment on "A New Jurisprudential Framework for Jurisdiction"' (2015) 109 AJIL Unbound 75.

Nagan, WP, 'Conflicts Theory in Conflict: A Systematic Appraisal of Traditional and Contemporary Theories' (1981–82) 3 J Intl & Comp L 343.

Nelson, Jason C, 'The United Nations and the Employment of Sanctions as a Tool of International Statecraft: Social Power Theory as a Predictor of Threat Theory Utility' (2005) 29 Law & Psychol Rev 105.

Nygh, PE, 'The Territorial Origin of English Private International Law' (1964) 2 U Tas L Rev 28.

Oster, Jan, 'Rethinking *Shevill*. Conceptualising the EU Private International Law of Internet Torts Against Personality Rights' (2012) 26 Intl Rev L, Compu & Tech 113.

Osula, 'Anna-Maria, Mutual Legal Assistance & Other Mechanisms for Accessing Extraterritorially Located Data' (2015) 9 Masaryk UJL & Tech 43.

Osula, Anna-Maria, and Mark Zoetekouw, 'The Notification Requirement in Transborder Remote Search and Seizure: Domestic and International Law Perspectives' (2017) 11(1) Masaryk UJL & Tech.

Parrish, Austen, 'The Effects Test: Extraterritoriality's Fifth Business' (2008) 61(5) Vanderbilt LR 1456.

Pauwelyn, Joost, 'The Use of Experts in WTO Dispute Settlement' (2002) 51 Intl & Comp LQ 325.

Post, David, 'Against "Against Cyberanarchy"' (2002) 17 Berkeley Tech LJ 1365.

Poullet, Yves, 'Transborder Data Flows and Extraterritoriality: the European Position' (2007) 2 J Intl Compu L & Tech 141.

Reidenberg, Joel R, 'Lex Informatica' (1998) 76(3) Tx L Rev 553.

Reidenberg, Joel R, 'Technology and Internet Jurisdiction' (2005) 153 U Pa L Rev 1951.

Reidenberg, Joel R, et al, 'Internet Jurisdiction: A Survey of Legal Scholarship Published in English and United States Case Law' (30 June 2013) Fordham Law Legal Studies Research Paper No 2309526 <http://ssrn.com/abstract=2309526> or <http://dx.doi.org/10.2139/ssrn.2309526>, accessed June 2017.

Revolidis, Ioannis, 'Judicial Jurisdiction over Internet Privacy Violations and the GDPR: a Case of "Privacy Tourism"?' 11(1) Masaryk UJL & Tech (2017) 7.

Rolph, D, 'Publication, Innocent Dissemination and the Internet after Dow Jones & Co Inc v Gutnick' (2010) 33 UNSWLJ 562.

Ryngaert, Cedric, 'An Urgent Suggestion to Pour Old Wine Into New Bottles: Comment on "A New Jurisprudential Framework for Jurisdiction"' (2015) 109 AJIL Unbound 81.

Ryston-Pratt, T, 'Olympic TV Rights' (2002) 21(4) Comm L Bull 12.

Schmitt, Michael N, 'In Defense of Due Diligence in Cyberspace' (22 June 2015) 125 Yale LJ Forum 68 <https://ssrn.com/abstract=2622077>, accessed June 2017.

Schultz, Thomas, 'Carving up the Internet: Jurisdiction, Legal Orders, and the Private/Public International Law Interface' (2008) 19 Eur J Intl L.

Schultz, Thomas, and Jason Mitchenson, 'Navigating Sovereignty and Transnational Commercial Law: The Use of Comity by Australian Courts' (2016) 12(2) J Priv Intl L 344–78 <http://dx.doi.org/10.1080/17441048.2016.1206704>, accessed June 2017.

Scott, Joanne, 'The New EU "Extraterritoriality"' (2014) 51 Common Mkt L Rev 1343.

Senz, Deborah, and Hilary Charlesworth, 'Building Blocks: Australia's Response to Foreign Extraterritorial Legislation' (2001) 2(1) Melb J Intl L 69.

Slane, A, 'Tales, Techs, and Territories: Private International Law, Globalization, and the Legal Construction of Borderlessness on the Internet' (2008) 71 L & Contemp Prob 129.

Slaughter, Anne-Marie, 'The Future of International Law Is Domestic (or, The European Way of Law)' (2006) 47 Harv Intl LJ 327.

Slaughter, Anne-Marie, 'A Liberal Theory of International Law' (2000) 94 Am Soc Intl L Proc 240.

Somek, Alexander, 'Kelsen Lives' (2007) 18 Eur J Intl L 409.

Stokes, M, 'Formalism, Realism, and the Concept of Law' (1994) 13 L & Phil 115.

Svantesson, Dan, 'What should Article 7—Consumer Contracts, of the Proposed Hague Convention, Aim to Accomplish in Relation to E-Commerce?' (September/October 2001) 17(5) Compu L & Sec Rep 318.

Svantesson, Dan, 'Geo-Location Technologies and Other Means of Placing Borders on the "Borderless" Internet' (Fall 2004) XXIII(1) John Marshall J Compu & Info L 101.

Svantesson, Dan, 'Borders on, or Border around—the Future of the Internet' (2006) 16(2) Albany LJ Sci & Tech 343.

Svantesson, Dan, 'How Does the Accuracy of Geo-location Technologies Affect the Law?' (2008) 2(1) Masaryk UJL & Tech 11.

Svantesson, Dan, 'The Choice of Courts Convention: How Will it Work in Relation to the Internet and E-Commerce?' (2009) 5(3) J Priv Intl L 517.

Svantesson, Dan, 'A Legal Method for Solving Issues of Internet Regulation; Applied to the Regulation of Cross-border Privacy Issues' (European University Institute Working Paper LAW2010/18).

Svantesson, Dan, 'Pammer and Hotel Alpenhof—ECJ Decision Creates Further Uncertainty about When E-Businesses "Direct Activities" to a Consumer's State under the Brussels I Regulation' (June 2011) 27(3) Computer L & Sec Rev 298–304.

Svantesson, Dan, 'Celebrating 20 Years of WWW—a Reflection on the Concept of Jurisdiction' (2012) 6(1) Masaryk UJL & Tech 177–90.

Svantesson, Dan, 'Time for the Law to take Internet Geo-location Technologies Seriously' (December 2012) 8(3) J Priv Intl L 473–87.

Svantesson, Dan, 'A "Layered Approach" to the Extraterritoriality of Data Privacy Laws' (2013) 3(4) Intl Data Priv L 278–86.

Svantesson, Dan, 'What is "Law", if "the Law" is Not Something That "Is"? A Modest Contribution to a Major Question' (September 2013) 26(3) Ratio Juris 456–85.

Svantesson, Dan, 'Between a Rock and a Hard Place—an International Law Perspective of the Difficult Position of Globally Active Internet Intermediaries' (2014) 30 Compu L & Sec Rev 348–56.

Svantesson, Dan, 'Delineating the Reach of Internet Intermediaries' Content Blocking—"ccTLD Blocking", "Strict Geo-location Blocking", or a "Country Lens Approach"?' (2014) 11(2) SCRIPT-ed 153–70.

Svantesson, Dan, 'The Extraterritoriality of EU Data Privacy Law—Its Theoretical Justification and Its Practical Effect on U.S. Businesses' (2014) 50(1) Stan J Intl L 53–102.

Svantesson, Dan, 'Sovereignty in International Law—How the Internet (Maybe) Changed Everything, but Not for Long' (2014) 8(1) Masaryk UJL & Tech 137–55.

Svantesson, Dan, 'A Jurisprudential Justification for Extraterritoriality in (Private) International Law' (2015) 13(2) Santa Clara J Intl L 517–71.

Svantesson, Dan, 'A New Jurisprudential Framework for Jurisdiction: Beyond the Harvard Draft' (2015) 109 AJIL Unbound 69 <https://www.asil.org/blogs/new-jurisprudential-framework-jurisdiction-beyond-harvard-draft>, accessed June 2017.

Svantesson, Dan, 'Extraterritoriality and Targeting in EU Data Privacy Law: The Weak Spot Undermining the Regulation' (2015) 5(4) Intl Data Priv L 226–34.

Svantesson, Dan, 'Limitless Borderless Forgetfulness? Limiting the Geographical Reach of the "Right to be Forgotten"' (2015) 2(2) Oslo L Rev 116–38.

Svantesson, Dan, 'The Relation between Public International Law and Private International Law in the Internet Context' (2006) 9 Yearbook of New Zealand Jurisprudence Special Edition 154–60.

Svantesson, Dan, 'The Holy Trinity of Legal Fictions Undermining the Application of Law to the Global Internet' (2015) 23(3) Intl JL & Info Tech 219–34.

Svantesson, Dan, 'Jurisdiction in 3D—"Scope of (Remedial) Jurisdiction" as a Third Dimension of Jurisdiction' (2016) 12(1) J Priv Intl L 60–76.

Svantesson, Dan, 'Against against Data Exceptionalism' (2016) 10(2) Masaryk UJL & Tech 200.

Svantesson, Dan, 'Article 4(1)(a) "Establishment of the Controller" in EU Data Privacy Law—Time to Rein In this Expanding Concept?' (2016) 6/3 Intl Data Privacy Law 210.

Svantesson, Dan, 'Nostradamus Lite—Selected Speculations as to the Future of Internet Jurisdiction' (2016) 10(1) Masaryk UJL & Tech 47.

Svantesson, Dan, and Felicity Gerry, 'Access to Extraterritorial Evidence: The Microsoft Cloud Case and Beyond' (2015) 31 Compu L & Sec Rev 478.

Svantesson, Dan, and Lodewijk van Zwieten, 'Law Enforcement Access to Evidence via Direct Contact with Cloud Providers—Identifying the Contours of a Solution' (2016) 32 Compu L & Sec Rev 671.

Trimble, Marketa, 'The Future of Cybertravel: Legal Implications of the Evasion of Geolocation' (2012) 22 Fordham Intellectual Property Media & Entertainment LJ 567.

Trimble, Marketa, 'Advancing National Intellectual Property Policies in a Transnational Context' (2014–15) 74 Md LR 203.

Van Alsenoy, Brendan, and Marieke Koekkoek, 'Internet and Jurisdiction after Google Spain: The Extraterritorial Reach of the "Right to be Delisted"' (2015) 5(2) Intl Data Priv L 105.

Von Glahn, Gerhard, 'Book Review: Répertoire suisse de droit international public: documentation concernant la pratique de la Confédération en matière de droit international public 1914-1939' (1978) 72 AJIL 197.

Weber, RH, 'Legal Interoperability as a Tool for Combating Fragmentation' (2014) Global Commission on Internet Governance Paper Series No 4 (Centre for International Governance Innovation).

Woods, AK, 'Against Data Exceptionalism' (2016) 68(4) Stan LR 729.

OTHER

Article 29 Working Group, 'Working Document on Determining the International Application of EU Data Protection Law to Personal Data Processing on the Internet by Non-EU Based Web Sites' (Working Paper 56, adopted on 30 May 2002).

Article 29 Working Party, *Opinion 2/2010 on online behavioural advertising* (22 June 2010) <http://ec.europa.eu/justice/policies/privacy/docs/wpdocs/2010/wp171_en.pdf>, accessed June 2017.

Article 29 Data Protection Working Party, 'Guidelines on the implementation of the Court of Justice of the European Union judgment on "Google Spain and Inc. v Agencia Española de Protección de Datos (AEPD) and Mario Costeja González" - C-131/12' (2014) WP225.

Barbour, EC, *The SPEECH Act: The Federal Response to 'Libel Tourism'* (Congressional Research Service, 2010) <http://www.fas.org/sgp/crs/misc/R41417.pdf>, accessed June 2017.

Barlow, John Perry, 'Jackboots on the Infobahn: Clipping the Wings of Freedom' (1993) 4 Wired <http://w2.eff.org/Misc/Publications/John_Perry_Barlow/HTML/infobahn_jackboots.html>, accessed June 2017.

Chatham House, 'The Principle of Non-Intervention in Contemporary International Law: Non-Interference in a State's Internal Affairs Used to Be a Rule of International Law: Is It Still?' (28 February 2007).

Commission Staff Working Document: A Digital Single Market Strategy for Europe—Analysis and Evidence Accompanying the Document Communication from the Commission to the European Parliament, the Council, the European Economic and Social Committee and the Committee of the Regions: A Digital Single Market Strategy for Europe {COM(2015) 192 final} Brussels, 6 May 2015.

Council conclusions on improving criminal justice in cyberspace (Luxembourg, 9 June 2016) <https://ec.europa.eu/home-affairs/sites/homeaffairs/files/what-we-do/policies/organized-crime-and-human-trafficking/council_conclusions_on_improving_criminal_justice_in_cyberspace_en.pdf>, accessed June 2017.

Edelman, B, 'Shortcomings and Challenges in the Restriction of Internet Retransmissions of Over-the-air Television Content to Canadian Internet Users' <http://cyber.law.harvard.edu/people/edelman/pubs/jump-091701.pdf>, accessed June 2017.

Fleischer, Peter, 'Response to the Questionnaire addressed to Search Engines by the Article 29 Working Party regarding the implementation of the CJEU judgment on the "right to be forgotten"' (31 July 2014) <https://docs.google.com/file/d/0B8syaai6SSfiT0EwRUFyOENqR3M/view?pli=1&sle=true>, accessed June 2017.

International Law Commission, *Report on the Work of its Fifty-Eighth Session* (1 May–9 June and 3 July–11 August 2006) UN Doc A/61/10.

Keller, Daphne, 'Global Right to Be Forgotten Delisting: Why CNIL is Wrong' (18 November 2016) <https://cyberlaw.stanford.edu/blog/2016/11/global-right-be-forgotten-delisting-why-cnil-wrong>, accessed June 2017.

Klein, S, 'Back to the 20th Century: The Olympics and the Internet' <http://mason.gmu.edu/~sklein1/index_files/contentexchangecolumns/column11.htm>, accessed June 2017.

Muir, JA, and PC van Oorschot, *Internet Geolocation and Evasion* (10 April 2006) <http://www.ccsl.carleton.ca/~jamuir/papers/TR-06-05.pdf>, accessed June 2017.

National Security Authority: Legislative Intent of the Law on Cybernetic Security (Czech Republic) <https://www.govcert.cz/download/legislativa/container-nodeid-660/legislative-intent-law-on-csen.pdf>, accessed June 2017.

'Right to be delisted' (Conseil d'État, 24 February 2017) <http://english.conseil-etat.fr/Activities/Press-releases/Right-to-be-delisted>, accessed June 2017.

Rogan, Aaron, 'US sues Microsoft for Silk Road suspect's email' (*The Times* [online], 22 June 2016) <https://www.thetimes.co.uk/article/us-sues-microsoft-for-silk-road-suspects-email-snq0tlqj0>, accessed June 2017.

Shavitt, Y, and N Zilberman, 'A study of geolocation databases' (2010) <http://arxiv.org/abs/1005.5674>, accessed June 2017.

Svantesson, Dan, 'Is Sharing the Answer to .BRAND Top Level Domain Disputes?' (CircleID, 17 June 2015) <http://www.circleid.com/posts/20150617_is_sharing_the_answer_to_brand_top_level_domain_disputes/>, accessed June 2017.

Swearing in Ceremony of the Honourable Justice Peter Hamill as a Judge of the Supreme Court of New South Wales (Tuesday 29 April 2104 [*sic*]) <http://www.austlii.edu.au/au/journals/NSWJSchol/2014/24.pdf>, accessed June 2017.

UN Human Rights Committee (HRC), *CCPR General Comment No. 16: Article 17 (Right to Privacy), The Right to Respect of Privacy, Family, Home and Correspondence, and Protection of Honour and Reputation*, 8 April 1988 <http://www.refworld.org/docid/453883f922.html>, accessed June 2017.

Westcott, Kathryn, 'Is King Canute misunderstood?' (BBC News, 26 May 2011) <http://www.bbc.co.uk/news/magazine-13524677>, accessed June 2017.

Whetstone, Rachel, 'Free expression and controversial content on the web' (14 November 2007) <http://googleblog.blogspot.com/2007/11/free-expression-and-controversial.html>, accessed June 2017.

Zakon, Robert H, 'Hobbes' Internet Timeline' <http://www.zakon.org/robert/internet/timeline/>, accessed June 2017.

Index

Printed and bound by CPI Group (UK) Ltd, Croydon, CR0 4YY